The Minerals, Metals & Materials Series

Bowen Li · Dipankar Ghosh · Eugene A. Olevsky ·
Kathy Lu · Faqin Dong · Jinhong Li ·
Ruigang Wang · Alexander D. Dupuy ·
Elisa Torresani
Editors

Advances in Powder and Ceramic Materials Science 2023

TMS

Springer

Editors
Bowen Li
Michigan Technological University
Houghton, MI, USA

Dipankar Ghosh
Old Dominion University
Norfolk, VA, USA

Eugene A. Olevsky
San Diego State University
San Diego, CA, USA

Kathy Lu
Virginia Polytechnic Institute and State University
Blacksburg, VA, USA

Faqin Dong
Southwest University of Science and Technology
Mianyang, China

Jinhong Li
China University of Geosciences
Beijing, China

Ruigang Wang
The University of Alabama
Tuscaloosa, AL, USA

Alexander D. Dupuy
University of California, Irvine
Irvine, CA, USA

Elisa Torresani
San Diego State University
San Diego, CA, USA

ISSN 2367-1181 ISSN 2367-1696 (electronic)
The Minerals, Metals & Materials Series
ISBN 978-3-031-22621-2 ISBN 978-3-031-22622-9 (eBook)
https://doi.org/10.1007/978-3-031-22622-9

Cover illustration: From Chapter "Fabrication of Ultra-Lightweight and Highly Porous Alumina Scaffolds by a Novel Sol–Gel/Freeze Casting Hybrid Method", Pei-Chieh Ho et al., Figure 3: SEM micrographs of alumina scaffolds in **a** in transverse section with 10vol% solid content **b** in longitudinal plane with 10vol% solid content **c** smooth lamellae surface of alumina scaffolds with nanoscale pores with 10vol% solid content **d** transverse section with 5vol% solid content **e** longitudinal plane with 5vol% solid content **f** discontinuous lamellae surface of alumina scaffolds with 10vol% solid content. https://doi.org/10.1007/978-3-031-22622-9_4.

This Springer imprint is published by the registered company Springer Nature Switzerland AG
The registered company address is: Gewerbestrasse 11, 6330 Cham, Switzerland

Preface

Ceramic materials science covers the science and technology of creating objects from inorganic, non-metallic materials, and includes design, synthesis, and fabrication of ceramics, glasses, advanced concretes, and ceramic-metal composites. Powder metallurgy and ceramics always share similar technical principals and methodologies in studying processes and structures of diverse materials from powder preparation, forming, sintering, until product finishing. In recent years, the hybrids of ceramic materials and metallic materials have received plenty of interdisciplinary inspirations and achievements in material processes and functional applications including ionic batteries, catalysis, energy storage, superconductors, semiconductor, filtrations, additive manufacturing, etc.

To promote interdisciplinary studies, The Minerals, Metals & Materials Society (TMS) Materials Characterization Committee and Powder Materials Committee have co-sponsored a symposium "Advances in Powder and Ceramic Materials Science" during the TMS Annual Meeting annually since 2020. This symposium is focused on the advances of powder and ceramic materials in the fundamental research, technology development, and industrial applications. The subjects of the symposium include synthesis, characterization, modeling, and simulation of powder and ceramic materials; design and control of ceramic microstructure and properties; ceramic powders and processing; surface treatment and thin films, membranes, and coatings of ceramics; hybrid systems of ceramic, metal, and/or polymer composites; and metallurgical byproducts for ceramic manufacturing. A special topic of "high-entropy ceramics" has been added to the 2023 symposium due to the superior properties of high-entropy ceramics with novel concepts that have attracted researchers to explore various potential structural and functional applications, while a number of materials are being re-acquainted and developed.

This volume of proceedings includes 16 reviewed manuscripts of original research from the symposia "Advances in Powder and Ceramic Material Science" and "Powder Materials Processing and Fundamental Understanding" (Sponsored by TMS Materials Processing & Manufacturing Division Powder Materials Committee). The manuscripts were invited or contributed by the researchers from the fields of powder processing, materials science, engineering, metallurgy, physics, manufacturing, and applications. This book provides authors' up-to-date achievements in powder and ceramic materials. It would be a useful reference for academic and industry readers.

The editors of this book would like to thank the authors for their contribution of manuscripts, the TMS Materials Characterization Committee and Powder Materials Committee for their support, and the TMS Program Committee and Springer for publishing this stand-alone volume.

Bowen Li
Lead Organizer

Contents

About the Editors

Bowen Li is a Research Professor in the Department of Materials Science and Engineering and Institute of Materials Processing at Michigan Technological University. His research interests include materials characterization and analysis, metals extraction, ceramic process, antimicrobial additives and surface treatment, porous materials, applied mineralogy, and solid waste reuse. He has published more than 140 technical papers in peer-reviewed journals and conference proceedings, authored/co-authored 3 books, and edited/co-edited 12 books. He also holds 16 patents and has delivered more than 30 invited technical talks. Dr. Li received a Ph.D. degree in Mineralogy and Petrology from China University of Geosciences Beijing in 1998 and a Ph.D. degree in Materials Science and Engineering from Michigan Technological University in 2008. He has been an active member in The Minerals, Metals & Materials Society (TMS), Society for Mining, Metallurgy & Exploration (SME), and China Ceramic Society. At TMS, he has served as the chair of the Materials Characterization Committee and as a member of TMS CDDC Committee, Powder Materials Committee, and the Biomaterials Committee. He has also served as an EPD Award committee member, *JOM* subject advisor, and key reader for *Metallurgical and Materials Transactions A*. He has been organizer/co-organizer of a number of international symposia and sessions. He also served as an editorial board member of the *Journal of Minerals and Materials Characterization and Engineering*, *Reviews on Advanced Materials Science*, and

FUTO Journal Series. He is the recipient of AIME Hal W. Hardinge Award (2022).

Dipankar Ghosh is currently an Associate Professor in the Mechanical and Aerospace Engineering Department at Old Dominion University (ODU), Norfolk, VA and joined the MAE faculty in July 2014. Prior to joining ODU, Dr. Ghosh was a postdoctoral scholar in the Division of Engineering and Applied Science at California Institute of Technology. He also held a position of postdoctoral associate in the Materials Science and Engineering Department at University of Florida. He completed his Ph.D. in Mechanical Engineering in 2009 from the University of Florida, Gainesville. Dr. Ghosh directs the Laboratory of Advanced Manufacturing of Materials. His research is focused on the freezing of particulate suspensions to develop novel directionally porous materials and multilayered composites. Dr. Ghosh is primarily working on advancing the freeze-casting technology for the synthesis and design of novel hierarchical ceramics and composites with a multidisciplinary research approach. Dr. Ghosh has published 2 book chapters, 47 peer-reviewed articles in prestigious journals, and is author/co-author in over 50 conference presentations. His research has been funded by the US National Science Foundation (NSF), Thomas F. and Kate Miller Jeffress Memorial Trust, Office of Naval Research (ONR), 4Va program, private industry, and internal grants from ODU Research Foundation.

Eugene A. Olevsky is the Dean and Distinguished Professor of the College of Engineering at San Diego State University. He has two M.S. degrees in Mechanical Engineering and Applied Mathematics and a Ph.D. in Materials Engineering. He is the author of over 500 publications and more than 150 invited presentations in sintering research. He has supervised more than 100 postdoctoral, graduate, and undergraduate students. Prof. Olevsky is a Fellow of the American Ceramic Society, a Fellow of the American Society of Mechanical Engineers, a Fellow of ASM International, Humboldt Fellow, and a Fellow of the International Institute of Science of Sintering.

Kathy Lu received her Ph.D. and M.S. degrees in Materials Science and Engineering from The Ohio State University, USA, and her B.S. in Ceramics from Tianjin University, China. From August 2004 to present, Dr. Lu has been a Professor in the Department of Materials Science and Engineering of Virginia Polytechnic Institute and State University, USA. During 2001–2004, she was a Materials Development and Processing Scientist at Energizer Battery Company, Westlake, Ohio, USA. During 2000–2001, she was the Director of Materials Processing, Center for Innovative Sintered Products of Pennsylvania State University, USA. She is the recipient of several awards including the 2022 TMS Fellow Award, 2020 Fulbright Distinguished Chair Award, 2017 Fellow of the American Ceramic Society, and 2015 Virginia Tech Alumni Award for Research Excellence.

Faqin Dong is a Professor in Applied Minerology at the Southwest University of Science and Technology, China. He has been devoted to the research and teaching of applied mineralogy, particuology, materials processing, and environmental protection for more than 40 years. He has conducted more than 40 research projects, published more than 400 papers, 7 edited books, held 19 patents, and supervised near 100 graduate students for their degree study. Currently, Dr. Dong serves as the President of the Southwest University of Science and Technology; Board Member of China Nonmetallic Mineral Industry Association; Vice President of China Mineralogy, Petrology and Geochemistry Society; Advisory Member of the National Natural Science Foundation of China; and Chairperson of International Congress of Applied Mineralogy Council. He has been involved in the organizing of many national and international conferences in Canada, Japan, Britain, Brazil, Czech Republic, Norway, and United States. He is the recipient of several awards including "Advanced Worker" by the State Ministry of Human Resources of China, "IncoPat Asian Inventor Award" by IncoPat Innovation Index Research Center, "Outstanding Young Teacher" by the Ministry of Education of China in 2002, and "Specialty Allowance" from the State Council of China.

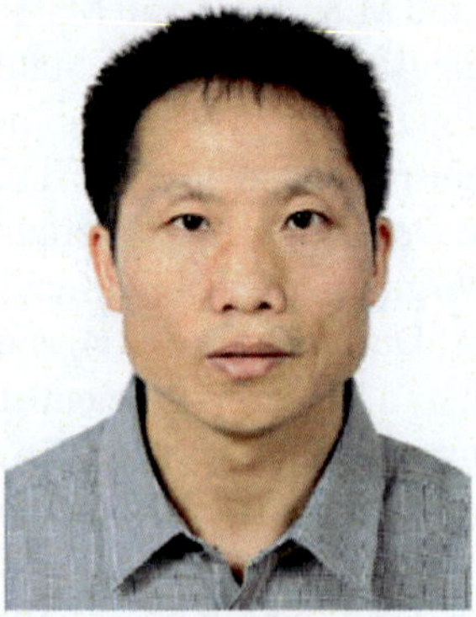

Jinhong Li is a Professor of the School of Materials Science and Engineering at China University of Geosciences, Beijing, China. He received his Ph.D. from China University of Geosciences, Beijing, China in 2007. His research interests include mineral energy storage materials, porous ceramics, and utilization of mineral and solid waste resource. He has undertaken more than 20 research projects; published more than 90 papers in international journals such as *Journal of Materials Chemistry A*, *Energy*, *Applied Energy*, *Journal of European Ceramic Society*, *Journal of Power Sources*, *Solar Energy Materials and Solar Cells*, and *Materials and Design;* co-authored two books; and served as a reviewer for a number of international journals. He also holds 6 patents. He was the winner of the New Century Talent Program from the Ministry of Education of China in 2008, the National Excellent Doctoral Dissertation in 2010, and was awarded the 12th Youth Geological Science Silver Hammer Award in 2011. He was a Visiting Scholar in the Department of Materials Science and Engineering at Pennsylvania State University, USA, from September 2011 to September 2012.

Ruigang Wang received his Ph.D. in Materials Science and Engineering from Arizona State University in 2007. His Ph.D. dissertation work was on in situ environmental transmission electron microscopy study of redox chemistry in CeO_2 and CeO_2-ZrO_2. He then did a postdoc at the Materials Sciences Division, Lawrence Berkeley National Lab, working on intermediate-temperature fuel cells and Li-ion battery materials. Currently, he is an Associate Professor in the Department of Metallurgical and Materials Engineering at The University of Alabama, Tuscaloosa. His research focuses on the synthesis and processing–structure–property relationship study of oxides, emission control catalysts, catalyst support materials, energy conversion/storage materials (fuel cell and battery, etc.), and high-temperature ceramic processing.

Alexander D. Dupuy is a project scientist in the Department of Materials Science and Engineering at the University of California, Irvine, where he studies the processing, properties, and phase transformation behavior of entropy stabilized oxides. He received his B.S. in Mechanical Engineering from the University of California Riverside (UCR) in 2009. His undergraduate research involved investigating gradient induced inhomogeneity found in the Spark Plasma Sintering (SPS) process. In 2011, he went on to receive his M.S. in Mechanical Engineering from UCR where he studied pressure induced densification mechanisms in nanomaterials processed using SPS. In 2016, he received his Ph.D. in Mechanical Engineering from UCR. His doctoral work focused on the processing and measurement of optical and ferroelectric ceramics.

Much of Dr. Dupuy's research has involved processing of functional and structural inorganic materials (ceramics, alloys, intermetallics, composites) with novel or customized phase and microstructural conditions, particularly involving nanostructuring and metastable states. He is particularly interested in the intersection between processing, microstructure, and phase transformations, and their influence on a wide range of material behaviors, including optical, ferroelectric, ferromagnetic, multiferroic, electrical, mechanical, and thermal properties. He also has a strong interest in materials with the potential for interesting phase transformation behavior, such as entropy stabilized oxides (entropic phase transformation) and ferroelectric materials (morphotropic/polymorphic phase boundaries), and how these transformations can be leveraged to enhance properties.

Elisa Torresani received her Ph.D. in Materials Science and Engineering from University of Trento (Italy) in 2016. In 2017, she started her postdoctoral research in the Department of Mechanical Engineering at San Diego State University, where she also taught a graduate level class on Mechanics of Sintering. Dr. Torresani's research is focused on understanding the fundamentals of sintering and developing constitutive models of sintering which enable the prediction of the macroscopic and microscopic behavior of porous bodies (shape distortion, density distribution, grain growth, etc.). Her current research emphasizes the coupling of experimental and theoretical investigations for a better understanding of the chemical and physical phenomena that drive and influence sintering-assisted additive manufacturing and field-assisted sintering processes. She has authored 19 refereed journal publications and has a patent. In 2022, one of her papers published in *Metallurgical and Materials Transactions A* received the AIME Champion H. Mathewson Award, which represents the most notable contribution to metallurgical science. Her postdoctoral work on the liquid phase sintering in microgravity has been supported by NASA.

Part I
Advances in Powder and Ceramic Materials Science

Design of High-Entropy Ceramics with IGZO-Based Compounds for Electroceramics Applications

Zaid Alejandro Luzanilla Meléndrez, Alejandro Durán, Francisco Brown, Ofelia Hernández Negrete, Javier Hernández Paredes, and Victor Emmanuel Alvarez Montano

Abstract Indium Gallium Zinc Oxide (IGZO) is a ceramic material used in optoelectronic technology, having essential results such as high efficiency in energy consumption and better image quality in display devices. On the other hand, high-entropy ceramics (HECs) are an important route to design new and novel functional materials, and whose physical properties need to be explored. In this work, it is studied the HECs with IGZO-based compounds synthesized using conventional solid-state method. Several trivalent and divalent cations are substituted in the layered crystal structure of IGZO, and the equilibrium of phases is studied under different compositions and high temperatures. The classic quenching method is used, cooling in air from 1400 °C to room temperature. Phase equilibrium was monitored using X-ray powder diffractometry (XRD) and scanning electron microscopy (SEM) to reveal microstructure and cation distribution with elemental mapping. In addition, once the single phase with high-entropy composition is obtained, the dielectric properties are explored to analyze their possible electroceramic applications.

Keywords High entropy ceramics · IGZO · Phase equilibrium

Z. A. Luzanilla Meléndrez · O. Hernández Negrete · V. E. Alvarez Montano (✉)
Departamento de Ingeniería Química y Metalurgia, Universidad de Sonora, Rosales y Luis Encinas s/n Col. Centro, Hermosillo, Sonora 83000, México
e-mail: victor.alvarez@unison.mx

A. Durán
Centro de Nanociencias y Nanotecnología, Universidad Nacional Autónoma de México, Km. 107 Carretera Tijuana-Ensenada, C.P 22860 Ensenada, Baja California, México

F. Brown
Departamento de Investigación en Polímeros y Materiales, Universidad de Sonora, Rosales y Luis Encinas s/n Col. Centro, Hermosillo, Sonora 83000, México

J. Hernández Paredes
Departamento de Física, Universidad de Sonora, Rosales y Luis Encinas s/n Col. Centro, Hermosillo, Sonora 83000, México

B. Li et al. (eds.), *Advances in Powder and Ceramic Materials Science 2023*, The Minerals, Metals & Materials Series, https://doi.org/10.1007/978-3-031-22622-9_1

Introduction

High-entropy ceramics (HECs) are materials in which at least four ions occupy a specific crystallographic site into crystal lattice forming a single solid solution [1]. These compounds are relatively new, discovered in the middle of the last decade, which have been shown to have interesting electrical, thermal, thermoelectric, and optical properties [2]. The possible combinations of elements to form HECs are so extensive that the search for new stable compounds that fall into this classification is currently on the rise, driven by the striking properties that they possess. In addition, being so relatively recent there is still much to explore regarding them.

On the other hand, IGZO-based compounds are the focus of research due to the wide capability that the hexagonal crystal structure has to substitute cations in the octahedral and trigonal bipyramidal coordination, as well as the homologous series that forms a c-axis aligned crystalline (CAAC) growing [3–6]. The compound $InGaZnO_4$ (IGZO) is a semiconductor oxide that has revolutionized the industry of flat screens and mobile devices; IGZO screens consume less energy, are cheap to manufacture, are resistant to weather conditions, and offer a sharpness and higher resolution compared to current liquid crystal displays [4, 5]. These kinds of materials are prepared not just as bulk but also as thin films under different conditions [7–10].

This work is focused on the design of HECs with IGZO-based compounds, in order to find the optimal conditions of synthesis that reveals a single phase, and thus investigate the possible potential applications that this interesting material can present. During this process, it was found that several conditions of temperature and composition lead to the equilibrium of the system towards the formation of a C-type, high entropic spinel, and high entropic layered IGZO type phases. Preliminary analysis of dielectric properties is presented.

Materials and Methods

Several mixtures of high purity oxides (>99.9%) with specific stoichiometry according to Table 1 were prepared following the conventional solid-state method. The oxide precursor in stoichiometric ratio was very well homogenized using agate mortar. Pellets of 13 mm diameter were made in a stainless-steel die applying a pressure of 2.5 ton. The pellets were placed in high alumina crucibles, and the solid-state reaction took place at high temperatures in the range of 1200 to 1400 °C with a subsequent classic quenching process in air in order to reach the desired single phase at the equilibrium. After each heating cycle, XRD analysis was made, until the equilibrium was reached. The equilibrium was considered when no change was observed in the XRD pattern between heating cycles. After that, SEM analysis and EDS spectra were obtained to verify the phase assembly. Specific samples were sintered at 1400 °C for 24 h with subsequent quenching to room temperature. For dielectric analysis, the sample was silver painted on the surface and cured in a furnace with a thermal

Table 1 Label of samples, stoichiometry of mixtures, heating treatments, and phases obtained at the equilibrium

Sample	Original mixture stoichiometry	Heating treatment	Final phases
M_1	$In(GaAlCrFe)_{0.25}(ZnNiMgCo)_{0.25}O_4$	1200 °C 1 day 1300 °C 2 days 1400 °C 2 days 1400 °C 2 days	C-type spinel
M_2	$In(GaCrFe)_{1/3}(ZnNiMgCo)_{0.25}O_4$	1200 °C 1 day 1300 °C 2 days 1400 °C 2 days 1400 °C 2 days	C-type spinel
M_3	$In(GaAlFe)_{1/3}(ZnNiMgCo)_{0.25}O_4$	1200 °C 1 day 1300 °C 2 days 1400 °C 2 days 1400 °C 2 days	C-type spinel layered
M_4	$InGa_{0.9}(AlCrFe)_{0.033}Zn_{0.9}(NiMgCo)_{0.033}O_4$	1400 °C 2 days	Layered

program from room temperature to 500 °C at a heating rate of 10/min. We use an LCR bridge (HP-4284A) to obtain the dielectric spectrum from room temperature to 550 °C in the frequency range of 1–50 kHz.

Results and Discussion

Table 1 resumes the information of the four samples prepared in this work, including the equilibrium final phases. Figure 1 shows the XRD pattern of sample M1 with its corresponding phase identification. Two phases are in equilibrium at the final state, cubic C-type and high entropic spinel. In the inset (Fig. 1), we can see the progress of the reaction, with temperature and time of heating. There, it can be seen that the stability of both phases is reached at 1400 °C. Figure 2 shows the Scanning Electron Microscopy image of sample M1 with its corresponding EDS spectra. Two phases can be observed in concordance with the XRD results, being the spinel the main

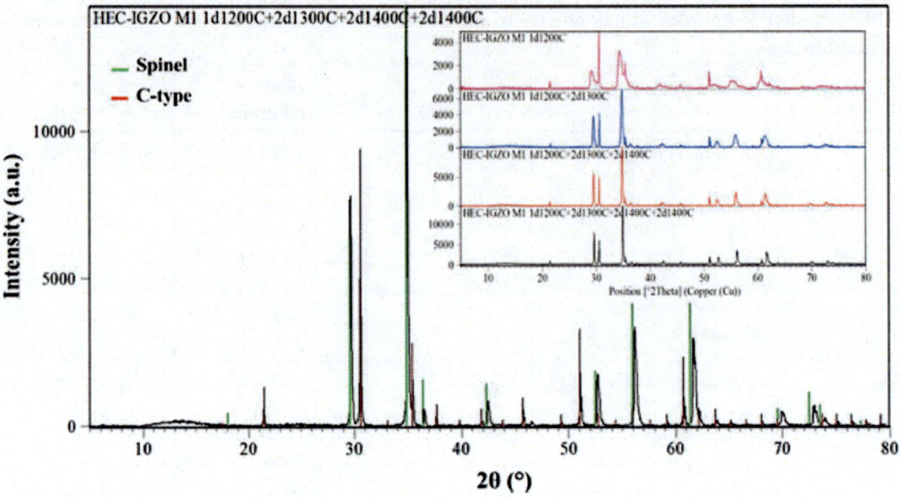

Fig. 1 XRD pattern and phase identification of sample M1 after several heating cycles. The inset shows the XRD patterns in order to observe the reaction progress after each cycle

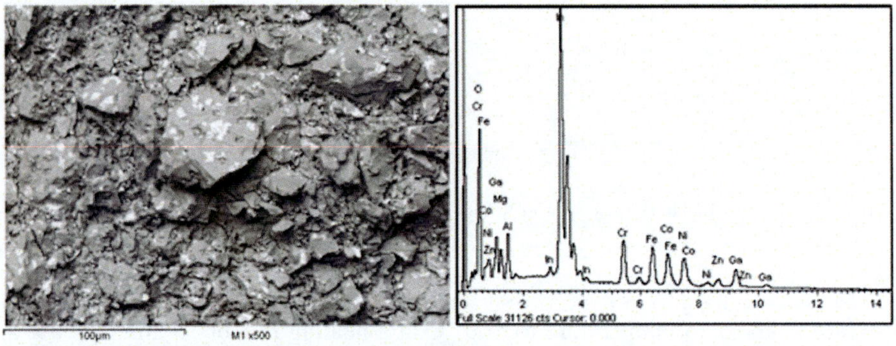

Fig. 2 SEM image and EDS spectra of sample M1 after several heating cycles

phase (gray color) and the cubic C-type observed as the white spot on the grains. All the cations introduced in the mixture can be found in the general EDS analysis. It is important to note that M1 sample represents the equimolar composition of all trivalent and divalent cations expected to introduce in the trigonal bipyramid site (CN = 5).

In Figs. 3 and 4, we present the XRD, SEM, and EDS spectra of samples M2 and M3, respectively. In these samples, aluminum or chromium was excluded in the specific stoichiometry. This was done in order to perturbate the system and to know what direction the reactions in the equilibrium will take. Figure 3 shows the same two phases that M1 sample, but small amount of C-type phase, was observed. The result infers that the spinel phase is of high entropy, and the C-type is indium-rich phase (see EDS spectra in the inset of Fig. 3). In the M3 sample, three phases occur

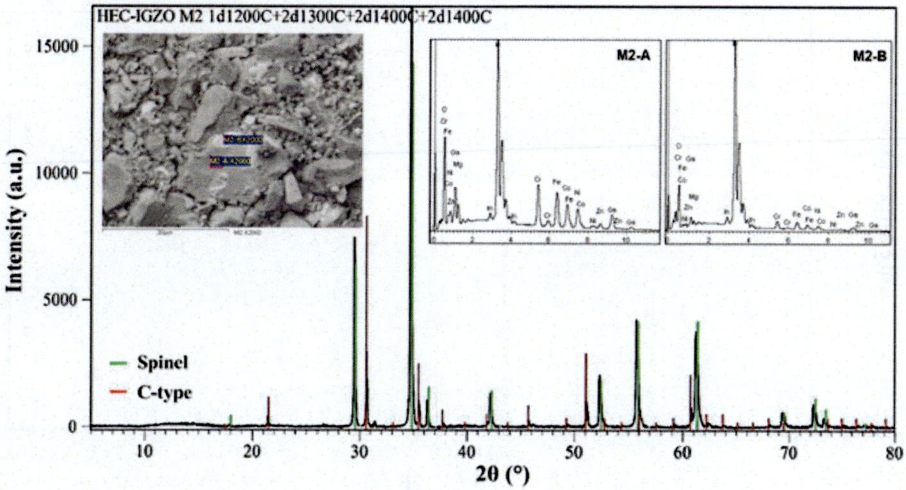

Fig. 3 XRD pattern and phase identification of sample M2 after several heating cycles. The inset figures show the SEM image and EDS spectra of the two observed phases

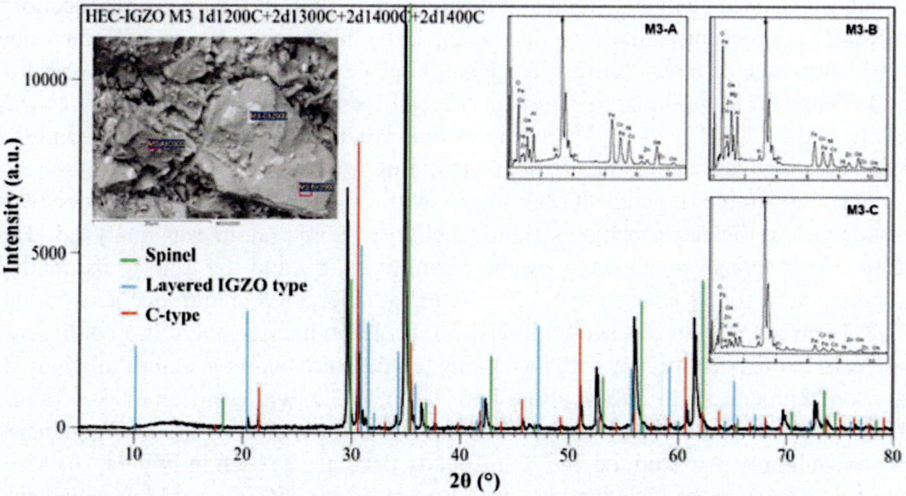

Fig. 4 XRD pattern and phase identification of sample M3 after several heating cycles. The inset figures show the SEM image and EDS spectra of the three observed phases

as seen in Fig. 4. Not only the spinel and cubic C-type phase, but also the layered IGZO type crystal structure begins to appear. The spinel and the layered IGZO type phases are high entropic while the C-type looks like just an In_2O_3 starting material phase.

In Fig. 5, the XRD pattern of M4 sample is presented. This sample has the characteristic that multicomponent trivalent and divalent cations are in very small amount in the IGZO stoichiometry (see Table 1). With this composition, the XRD result shows

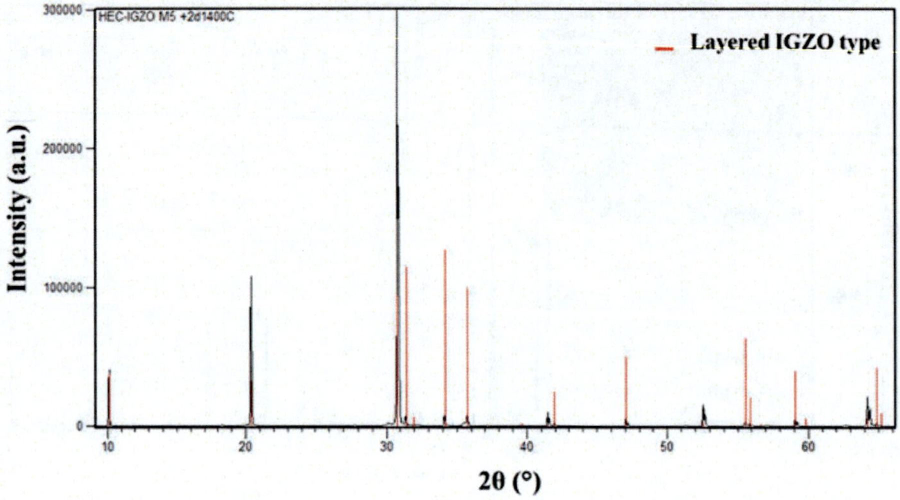

Fig. 5 XRD pattern and phase identification of sample M4 after a single heating cycle

a single phase with a closely related crystal structure to IGZO. Almost all reflections of IGZO are present, and as can be expected, the intensities are very different, due to the presence of many cations in the single phase. In Fig. 6, SEM, EDS spectra, and Elemental Mapping are presented. We can see that single phase occurs as was confirmed by XRD pattern. There, the cations are homogenously distributed in the powder sample as seen in the elemental mapping images.

Once a simple layered IGZO phase with high-entropy characteristics was obtained, the dielectric behavior as a function of temperature was analyzed. The dielectric response represented by the permittivity, ε', and the Tan δ, also called dissipation factor, as a function of temperature at several selected frequencies for the IGZO-entropic ceramic is shown in Fig. 7a, b). In both cases, there is a continuous increase in the permittivity, with increasing temperature taking ε' values of about 80 at room temperature to 25 k at about 550 °C for 1 kHz, while the tan δ takes values from ~0.87 at room temperature to ~280 at 550 °C for 1 kHz. At higher frequencies is shown that both ε' and the Tan δ drastically decrease as seen in Fig. 7a, b). Also, we can observe in the lost tangent a peak formation at ~450 °C, and low frequencies that disappear at higher frequencies (Fig. 7b). It is known that the IGZO compound is a semiconductor with a high electronic mobility [7]. The high permittivity values could be related to two factors (i) the high electronic mobility observed in the IGZO or (b) the disorder associated with the cations in the IGZO matrix. A comparative study of matrix and entropic compound is in progress in order to elucidate the true effect of multicomponent cations in an IGZO-based compound.

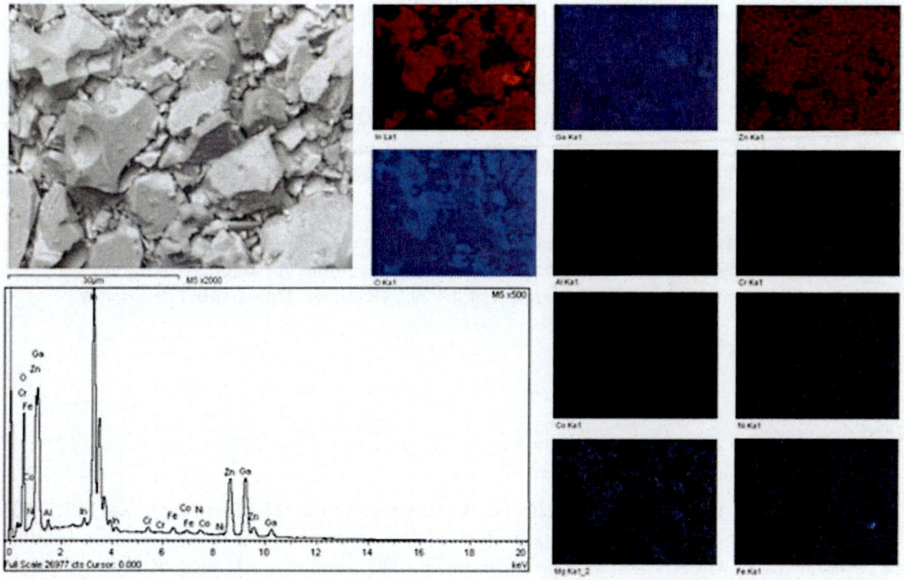

Fig. 6 SEM image, EDS spectra, and Elemental Mapping analyses of sample M4 in powder form sintered at 1400 °C

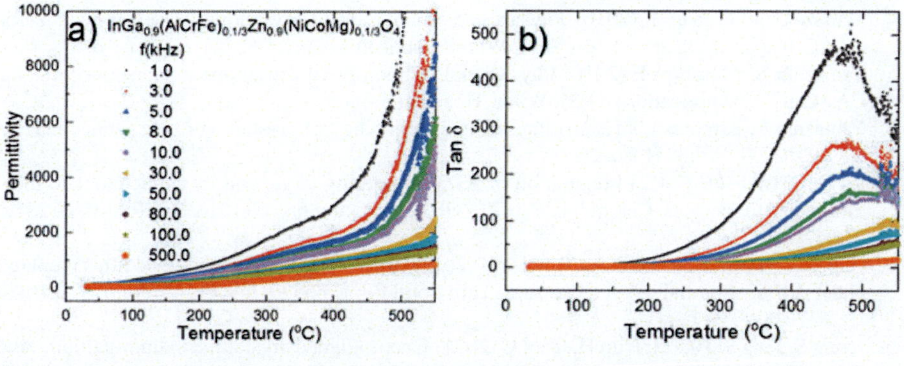

Fig. 7 Temperature-dependence of dielectric permittivity, ε′ and lost tangent, Tan δ at several frequencies for $InGa_{0.9}(AlCrFe)_{0.03}Zn_{0.9}(NiCoMg)_{0.03}O_4$ entropic ceramic

Conclusion

The phase equilibrium of several compositions with IGZO-based stoichiometry using the HECs approach was studied with XRD and SEM–EDS analysis. The different compositions studied showed that the system tends to the formation of two or three phases in the equilibrium, to say C-type, spinel, and layered IGZO type phase. We found that the equilibrium can be modified toward the spinel or layered phase, with variation of the cations incorporated. In a highly closed IGZO composition with small amounts of cations, $InGa_{0.9}(AlCrFe)_{0.03}Zn_{0.9}(NiCoMg)_{0.03}O_4$, a single high-entropy phase was stabilized with layered IGZO type structure. A preliminary dielectric analysis on this layered IGZO ceramic compound showed a continuous increase in both the ε' and Tan δ with increasing temperature from room temperature to ~550 °C for 1 kHz. More analyses are in progress to elucidate the possible applications as electroceramic devices.

Acknowledgements ZALM acknowledges the support from CONACyT for his graduate studies. AD and VEAM appreciate the support through the 'Intercambio Académico' program promoted by the Coordination of Scientific Research (CIC)-UNAM and the Universidad de Sonora.

References

1. Oses C, Toher C, Curtarolo S (2020) High-entropy ceramics. Nat Rev Mater 5:4.5:295–309. https://doi.org/10.1038/s41578-019-0170-8
2. Akrami S, Edalati P, Fuji M, Edalati K (2021) High-entropy ceramics: Review of principles, production and applications. Mater Sci Eng R Rep 146:100644. https://doi.org/10.1016/J.MSER.2021.100644
3. Eichhorn S, Schmid H, Assenmacher W, Mader W (2017) Homologous compounds of type $ARO_3(ZnO)m$ in the system Ga–Sn–Zn–O. J Solid State Chem 246:214–220. https://doi.org/10.1016/J.JSSC.2016.11.031
4. Yamazaki S, Tsutsui T (2016) Physics and technology of crystalline oxide semiconductor CAAC-IGZO: application to displays. Wiley, Hoboken
5. Yamazaki S, Masahiro F (2016) Physics and technology of crystalline oxide semiconductor CAAC-IGZO: application to LSI. Wiley, Hoboken
6. Yamazaki S, Kimizuka N (2016) Physics and technology of crystalline oxide semiconductor CAAC-IGZO. Wiley, Hoboken
7. Lo CC, Hsieh TE (2012) Preparation of IGZO sputtering target and its applications to thin-film transistor devices. Ceram Int 38:3977–3983. https://doi.org/10.1016/J.CERAMINT.2012.01.052
8. Nagata T, Hiraoka K, Okamoto T, Iwata N (2018) Crystal structure of YbFe2O4 films prepared under different partial oxygen pressure. Thin Solid Films 665:96–98. https://doi.org/10.1016/J.TSF.2018.09.012
9. Jeong S, Jang S, Han H, Kim H, Choi C (2021) C-axis aligned crystalline indium-gallium-zinc oxide (CAAC-IGZO) and high-k charge trapping film for flash memory application. J Alloys Compd 888:161440. https://doi.org/10.1016/J.JALLCOM.2021.161440
10. Nomura K, Ohta H, Ueda K, Kamiya T, Hirano M, Hosono H (1979) Thin-film transistor fabricated in single-crystalline transparent oxide semiconductor. Science 300(2003):1269–1272. https://doi.org/10.1126/SCIENCE.1083212

Development of High Voltage Multilayer Ceramic Capacitor

Hyungsuk K. D. Kim

Abstract MLCC serves as a filter for reducing the noise of electric current. $BaTiO_3$ (BT) is currently used as a dielectric material for high capacitance MLCCs. As power and charging speed of the EVs improve, higher voltages need to be selected. BT has material limitation in that the dielectric constant decreases with increasing voltage. Therefore, sacrifices have to be made in capacitance by decreasing permittivity or antiferroelectric materials need to be utilized. In this study, PLZT-based MLCC with higher dielectric constant than BT in 200–400 V was developed. The contents of Pb, La, Zr, and Ti were optimized, and Ni electrode was replaced with Cu to reduce cost. To solve possible Cu oxidation and PbO reduction, oxygen partial pressure was optimized.

Keywords High capacitance MLCC · $BaTiO_3$ · PLZT

Introduction

$BaTiO_3$ (BT) is the most representative material for dielectrics used in MLCCs [1, 2]. BT is ferroelectric below 120 °C, and the dielectric constant and capacitance of it decrease as the voltage applied increases. Recently, as the voltage of the battery and charging system consistently increase in order to enhance the charging speed of the battery in electric vehicles, the capacitance of MLCC at the applied voltage becomes smaller and needs to be improved. It has been suggested as a possible candidate to use antiferroelectrics, whose dielectric constant and capacitance increase with the voltage, to the MLCC. It may lead to system miniaturization and cost reduction by reducing the number of MLCCs used by replacing MLCCs with lower capacitance.

H. K. D. Kim (✉)
Hyundai Motor Company, 37, Cheoldobangmulgwan-Ro, Uiwang-Si, Gyeonggi-Do, Republic of Korea
e-mail: hssaga@hyundai.com

B. Li et al. (eds.), *Advances in Powder and Ceramic Materials Science 2023*, The Minerals, Metals & Materials Series, https://doi.org/10.1007/978-3-031-22622-9_2

Material Development

Optimization of PLZT Composition

The composition of PLZT, the prototypical antiferroelectric material, was optimized. Pb exists in the form of PbO in the material and PbO can be reduced to Pb when sintered in a reducing atmosphere. To prevent this, the amount of PbO was compensated by adding excess PbO. It was confirmed that the dielectric characteristics varied depending on La and Ti contents. The contents of La and Ti were optimized so that the characteristics of the PLZT show antiferroelectric. As shown in Figs. 1 and 2, the dielectric characteristics were changed according to the contents, and in terms of dielectric constant, it tended to decrease as the La content increased. As the content of Ti increased, the dielectric constant tended to improve. As a result, a PLZT material with a high dielectric constant was made with the contents of La and Ti being 8% and 14%, respectively.

In terms of material development, efforts were made to reduce the sintering temperature besides composition optimization. While Ni normally is used as an electrode in the MLCC with BT as a dielectric material, Cu is used as an electrode to reduce cost in this study. Considering the fact that the sintering temperature of a specific PLZT composition that we developed is approximately 1,100 °C and the melting point of Cu is 1,080 °C, it was necessary for the sintering temperature of MLCC to be less than 1,000 °C. For this purpose, the particle size of PLZT powder

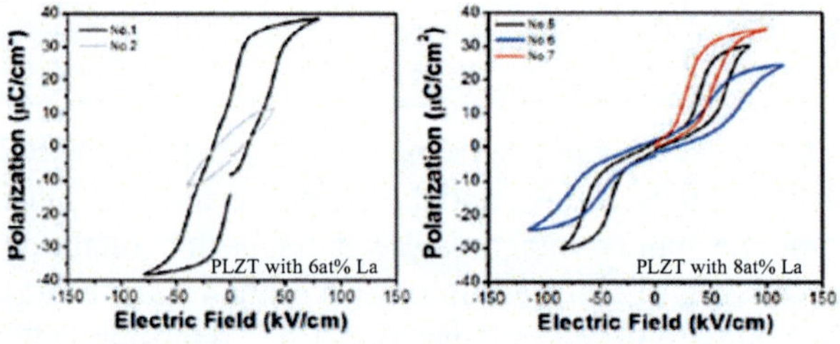

Fig. 1 PE curve of PLZT depending on La content

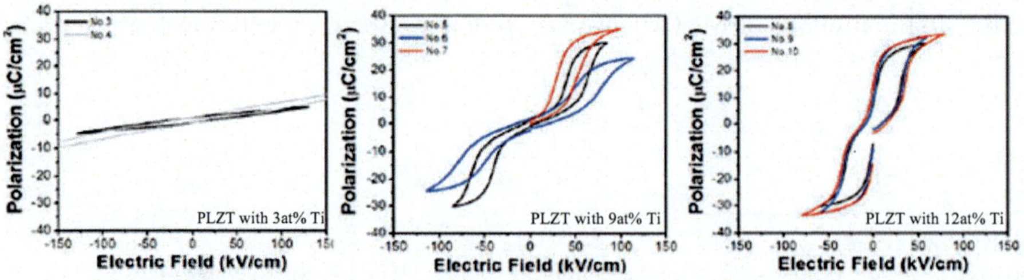

Fig. 2 PE curve of PLZT depending on Ti content

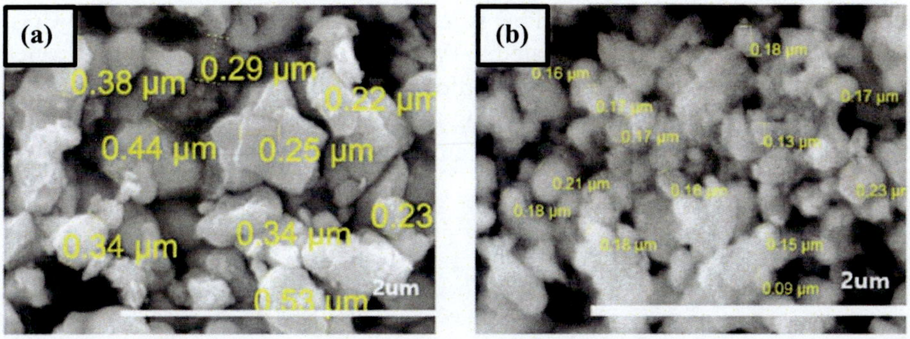

Fig. 3 Size comparison of PLZT powder before **a** and after **b** uniformization

Fig. 4 Effect of low-temperature sinterable glasses on grain size and uniformity

was uniformized (D50: 0.5–0.3 um, Fig. 3) by making PLZT powder smaller, which leads to lowering of the sintering temperature due to the increase of surface area of the powder.

As an additional method for reducing the sintering temperature, two types of low-temperature sinterable glasses were investigated: Pb-based glass and Li-based glass. The low-temperature sinterable glass melts below the sintering temperature, so that both high density through liquid sintering and lower sintering temperature can be achieved. The glass transition temperature (Tg) of Pb-based glass is higher than the Li-based glass; thus, there is no possibility of blocking the passage of the binder during burnout. In the end, the Pb-based glass was chosen due to more uniform grain size and no secondary phase formed with the base material as depicted in Fig. 4. Additionally, Pb-based glasses can play a role of excess PbO to compensate PbO volatility.

Optimization of PLZT Heat Treatment Process

The atmosphere of the heat treatment process of PLZT requires temperature and oxygen partial pressure control. While proper oxygen partial pressure should be maintained for each temperature, PbO is reduced to Pb ($PbO + CO \rightarrow Pb + CO_2$) when oxygen is insufficient compared to the proper oxygen partial pressure, which leads to degradation of dielectric properties of PLZT. When oxygen is excessive, the Cu electrode is oxidized; thus, during the degassing and sintering process, the oxygen partial pressure and temperature are required to be controlled.

Test Results

The MLCC samples having two different specifications were manufactured so that their maximum capacitance is achieved at 250 and 350 V, respectively, for batteries in hybrid electric vehicles. The size is 5750 and the thickness is 1.4 T (5.7 * 5 * 1.4 mm), and the specifications for capacitance at different voltages are 4uF @ 200 V, 3uF @ 300 V, and 2uF @ 400 V. In terms of how MLCCs are arranged, four MLCCs are aligned vertically in one metal frame as shown in Fig. 5. The characteristics of the MLCC are shown in Table 1. It was interesting to note that the voltage from which the maximum capacitance is generated is proportional to the thickness of dielectric material to a certain extent.

The capacitance of the developed MLCC was measured depending on different voltages as depicted in Fig. 6. It was confirmed that each MLCC showed maximum

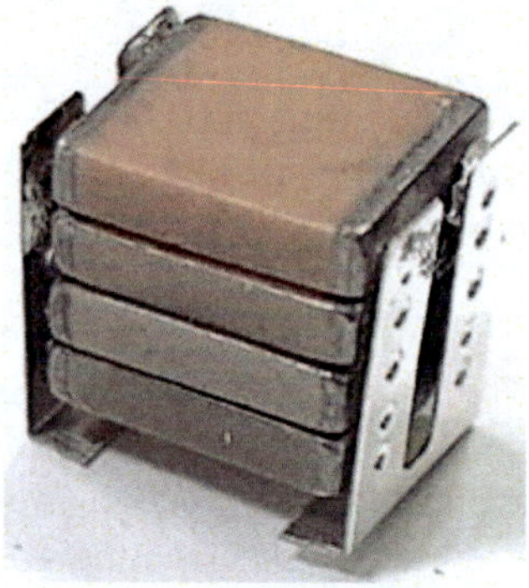

Fig. 5 MLCC sample configuration

Table 1 Characteristics of developed MLCC samples

	Operating voltage [V]	Co [nF]	Ceff [nF]	BDV [V]	Thickness of dielectrics [um]	Grain size [nm]	Permittivity
Spec #1	250 V	500	900	500	20	1,000–1,500	1,100
Spec #2	350 V	300	600	750	30		

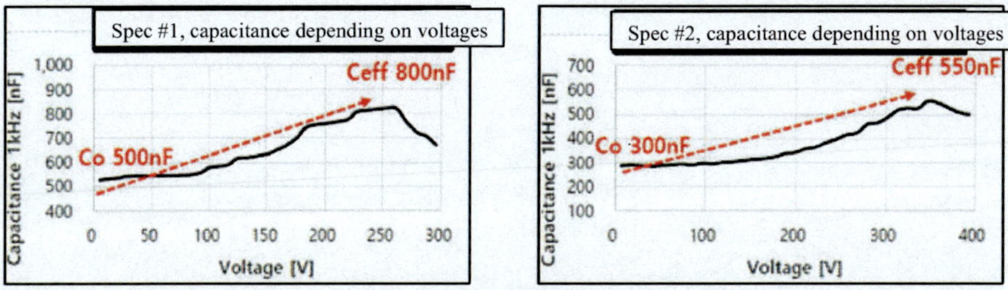

Fig. 6 Capacitance measurement results depending on voltages

Table 2 Number of MLCC required to satisfy capacitance at each voltage

Specification		250 V 用	350 V 用
4000 nF	@ 200 V	750 * 4 * 2 = 6000	350 * 4 * 3 = 4200
3000 nF	@ 300 V	650 * 4 * 2 = 5200	500 * 4 * 2 = 4000
2000 nF	@ 400 V	-	500 * 4 * 1 = 2000

capacitance at the desired voltages as designed. Currently, four metal framed, BT-based MLCCs (4 MLCCs in 1 metal frame) are used for both specifications. As shown in Table 2, number of metal framed MLCCs can be reduced to two for 250 V specifications and to three metal frames for 350 V specifications when using the PLZT-based configuration as shown in Table 2.

Conclusion

MLCCs with maximum capacitance at 250 and 350 V for batteries in hybrid electric vehicles were developed. It was found that the voltage at which the maximum capacitance is realized is proportional to the thickness of dielectrics, and it was confirmed that the number of MLCCs can be reduced compared to the existing BT-based MLCCs because the capacitance was improved.

References

1. Hong K et al (2019) Perspectives and challenges in multilayer ceramic capacitors for next generation electronics. J Mater Chem C 7.32:9782–9802
2. Pithan C, Hennings D, Waser R (2005) Progress in the synthesis of nanocrystalline $BaTiO_3$ powders for MLCC. Int J Appl Ceram Technol 2(1):1–14

Development of an Experimentally Derived Model for Molybdenum Carbide (Mo_2C) Synthesis in a Fluidized-Bed Reactor

Maureen P. Chorney, Jerome P. Downey, and K. V. Sudhakar

Abstract Experiments were conducted to evaluate molybdenum carbide, Mo_2C, synthesis in a fluidized-bed reactor. Molybdenum was introduced to the reactor as a precursor formed by adsorption of molybdate ions on an activated carbon substrate. Design of experiments was accomplished through the use of commercial software, Design-Expert12®. A matrix of seventeen experiments was developed and completed to evaluate molybdenum carbide synthesis as a function of reaction time, reaction temperature, and reactive gas composition. Conversion efficiencies were determined by characterizing the experimental products via X-ray Diffraction (XRD) and Scanning Electron Microscopy (SEM). The conversion model was created through the application of response surface methodology utilizing a central composite design. Confirmatory experiments were performed to validate the model.

Keywords Fluidized bed · Carbide synthesis · Modeling

Introduction

Molybdenum carbide (Mo_2C), like other transition metal carbides, is a sought-after material for its exceptional chemical, physical, and mechanical properties. Similar to tungsten carbide, Mo_2C is utilized in many high temperature, thin film, and cutting tool applications [1–3]. Additionally, the crystal structure of molybdenum carbide leads to a contraction of the *d*-band, resulting in an increase in electron density that provides Mo_2C with desirable electronic, electrochemical, and catalytic properties [1]. Many catalytic processes rely on noble metals, and substitutes, such as Mo_2C, can provide more economical and stable catalyst opportunities [4–7].

Molybdenum carbide synthesis can be simply denoted as the carbothermal reduction of molybdenum oxide (MoO_3 or MoO_2) in the presence of carbon to form molybdenum carbide (Mo_2C) and carbon monoxide (Eqs. 1 and 2) [2, 8].

M. P. Chorney (✉) · J. P. Downey · K. V. Sudhakar
Montana Technological University, Butte, MT 59701, USA
e-mail: mchorney@mtech.edu

B. Li et al. (eds.), *Advances in Powder and Ceramic Materials Science 2023*, The Minerals, Metals & Materials Series, https://doi.org/10.1007/978-3-031-22622-9_3

$$2MoO_3 + 7C = Mo_2C + 6CO \quad (1)$$

$$2MoO_2 + 5C = Mo_2C + 4CO \quad (2)$$

The conversion of molybdenum oxide to carbide occurs as a multistep process, with the oxide first reducing to metallic molybdenum, reacting with carbon to form MoC, and then on to form Mo_2C as function of time and temperature [2]. By creating an even more advantageous reaction atmosphere through the addition of methane (CH_4), additional pathways to produce molybdenum carbide are presented. Nayak et al. present several equations that illustrate the reactions that may occur in the presence of CH_4 (Eqs. 3 and 4) and the resultant hydrogen produced from cracking methane [9].

$$2Mo + CH_4 = Mo_2C + 2H_2 \quad (3)$$

$$2MoC + 4H_2 + C = Mo_2C + 2CH_4 \quad (4)$$

Fluidization technology provides the superior gas–solid contact, compared to the more limited contact of tube and rotary furnaces [10], needed for efficient synthesis. Fluidized-bed technology provides excellent heat and mass transfer characteristics, which improve solid–gas reactions [11–13]. A more uniform reaction temperature and atmosphere are achieved in fluid bed reactors, and they can be easily scaled to the desired throughput requirements [11, 12].

Materials and Methods

Precursor Preparation

A molybdenum precursor was prepared by adsorbing molybdate anions (sodium molybdate dihydrate, Sigma Aldrich, ≥99.5%) onto an activated charcoal substrate (Sigma Aldrich, untreated, 100–400 mesh). A solution containing 18,000 ppm Mo and 0.2 M NaCl (Sigma Aldrich, anhydrous, ReagentPlus®, ≥99%) was adjusted to pH 2.0, prior to the introduction of activated carbon. The solution was agitated continuously after the addition of carbon for 2.0 h, and then, the precursor was vacuum filtered and placed in a drying oven.

Fluidized Bed

All scoping and design matrix experiments utilized a vertical MTI 1200X fluidized-bed furnace. Experiments were performed in a quartz tube with a 22 mm inner diameter bed region and a 44 mm freeboard. A porous quartz frit keeps sample in place within the tube, while allowing gas flow to interact with the fixed bed of particles to create a fluidized bed. Each experiment was conducted with approximately 2.0 g of precursor, which is the maximum amount of sample that allows for smooth fluidization, rather than slugging or channeling. Given the small surface area of the fluidized region, as well as the fine particle size, the gas flow rate for each experiment was 0.6 LPM.

Design of Experiments

A design matrix was developed in Design-Expert12® following a full central composite response surface design with three midpoints. Three factors—reaction time, temperature, and gas composition—were varied over the course of seventeen experiments. High and low values were determined based on previous scoping experiments. For each experiment, argon was used to purge the tube to maintain an inert atmosphere as the furnace was heated to temperature at a 5.3 °C/min average ramp rate. Argon flow was also important to keep particles fluidized to prevent agglomeration of the precursor that could inhibit fluidization during the hold at temperature. Once the reaction temperature was reached, the gas mixture—methane and carbon monoxide—was introduced to the furnace using mass flow controllers to tightly control the composition of the inlet gas. After the completion of the experiment, the furnace was set to cool at a rate of 3 °C/min to 400 °C, from which it was allowed to cool naturally, under argon gas flow.

Characterization

Characterization of experiment products was performed primarily using a Rigaku Ultima IV X-ray diffractometer (Cu Kα source) and SmartLab Studio II analysis software. Imaging of samples was conducted using a TESCAN MIRA3 scanning electron microscope (SEM). Design-Expert12® (Stat-Ease) software was used to perform the statistical analysis of the design matrix.

Results

Synthesis of Molybdenum Carbide

Scoping experiments were completed and analyzed with SmartLab Studio II software to establish design matrix values. Temperature, hold time, and reaction gas composition were the three design factors utilized, while gas flow rate at temperature was held constant to maintain smooth fluidization throughout the experiments. Table 1 provides representative WPPF analysis results of corresponding XRD diffractograms. Each response value is the average of three separate diffractograms for each experiment. Figure 1 illustrates the differences between diffractograms of the high and the low temperature experiments, and Fig. 2 provides SEM imaging of a center design point experiment.

Table 1 Select results from design matrix for the synthesis of molybdenum carbide (Mo_2C)

Factors			Response	
A-Temp. (°C)	B-Time (Hr.)	C-Methane (%)	Mo_2C (%)	MoC (%)
800	6	70	69.7	30.3
880	6	70	97.0	3.0
880	6	90	94.0	6.0
840	4	80	94.2	5.8
893	4	80	99.4	0.6

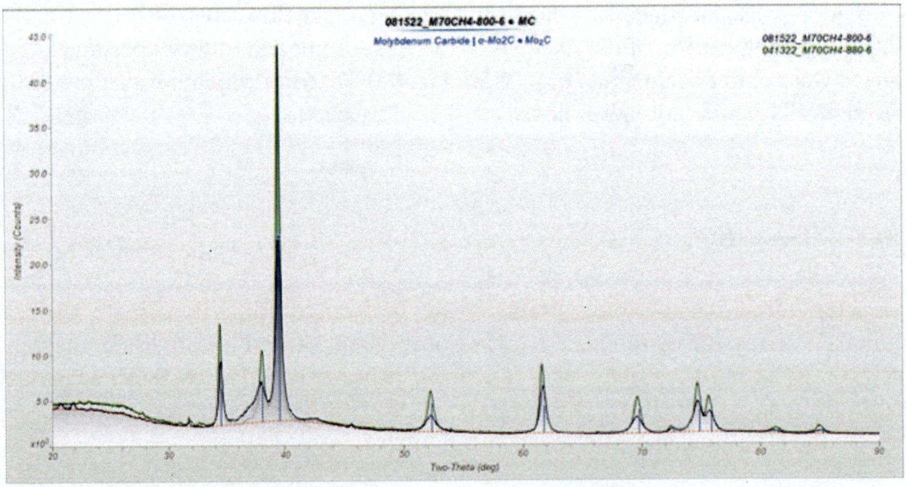

Fig. 1 Diffractogram overlay of 70% CH_4—880 °C—6 h. (green, back) and 70% CH_4—800 °C—6 h. experiments (blue, front)

Fig. 2 SEM image of experiment conducted in an 80% methane atmosphere at 840 °C for 4 h. Submicron particles are observed on the activated carbon substrate

Several trends can be observed from the select results presented in Table 1. Firstly, temperatures below 840 °C are not as successful in fully converting to Mo_2C. Secondly, more carbon monoxide present in the gas mixture appears to be beneficial to Mo_2C conversion, with a 3.0% increase noted for the 880 °C experiments. Additionally, increasing the temperature by 13 °C above the original design space results in almost complete conversion to Mo_2C, while using 80% methane gas (mid-value) for the reaction atmosphere.

The XRD pattern shows an increase in peak intensity as reaction temperature increases. The increase in peak intensity indicates that more molybdenum carbide (Mo_2C) is present at higher temperatures. Also, it is observed that peaks are generally broader at lower temperatures, while no additional peaks are observed. A broader peak may indicate that those experiments conducted at lower temperatures result in a product with less order, in terms of crystal structure, than those performed at higher temperatures, which show increased order based on peak width. When analyzing the same three diffractograms of the 800 °C, 70% CH_4, and 6 h experiment using JADE software, 98.6% Mo_2C was found, compared to 69.7% Mo_2C reported by SmartLab Studio II, as shown in Table 1. SmartLab Studio II software seems to be more sensitive to short range vs. long range order of the sample, as observed by the increase in MoC calculated during the WPPF analysis (800 °C, 70% CH_4, and 6 h experiment), when no additional peaks of great intensity are observed for MoC [14].

Modeling of Molybdenum Carbide Synthesis

A quadratic model best fits the Mo_2C synthesis data, with temperature and gas composition as key factors, as well as the interaction between these two variables. The

F-value of the model, 33.60, indicated that the model was significant, and there is less than 0.0001% chance that this F-value was due to noise. The model terms—temperature (A), p-value <0.0001; AC, p-value 0.0071; and A^2, p-value 0.0018—were deemed to be significant with p-values less than 0.0050. Methane (C), p-value 0.0743, was deemed not significant, but was included in the model due to the significance of the interaction between gas composition and temperature. The residual lack of fit was not significant with a p-value of 0.1254.

The R^2 of the model was found to be 0.9180, and the adjusted R^2 and predicted R^2 were 0.8907 and 0.8684, respectively, which are in reasonable agreement. The adequate precision of the model, which measures the signal-to-noise ratio, was 16.07 (must be greater than 4 to be considered a desirable signal-to-noise ratio). The model was deemed to be a good fit for the data. Figures 3 and 4 provide the model graph data.

The response surface, Fig. 3, provides a visual means of evaluating the best parameters for the conversion to Mo_2C. Gas composition is an important variable, with methane percentages less than 75% resulting in better conversion; additionally, higher temperatures, greater than 860 °C, also relate to better Mo_2C conversion. The design points, designated as red circles, are the three center points of the matrix—840 °C, 80% CH_4, and 4 h—and are found to sit slightly above the response surface. Figure 4 provides additional plots to validate the model; and the model equation which may be used to determine percent Mo_2C, in terms of the actual factors in their original

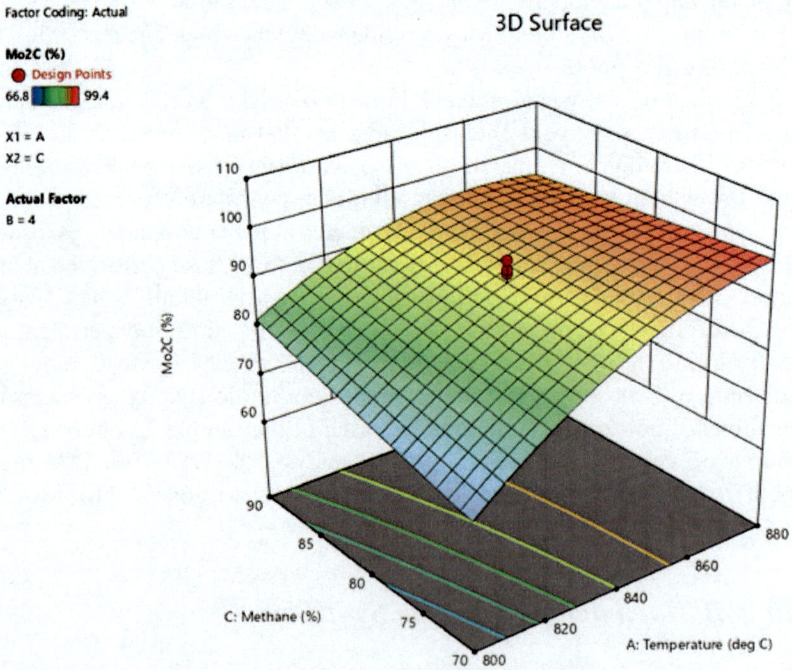

Fig. 3: 3D response surface of the synthesis of molybdenum carbide in a fluidized-bed reactor

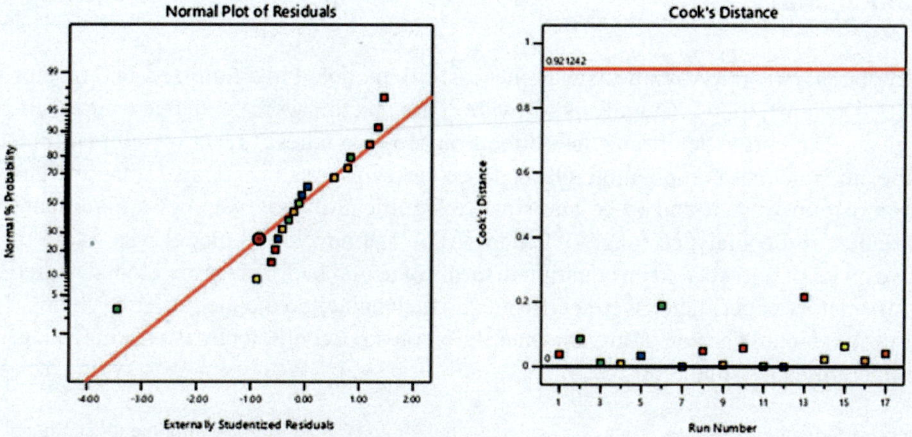

Fig. 4 Normal plot of residuals (left) and Cook's distance plot (right) for the synthesis of Mo_2C

units (Table 1), is presented below (Eq. 5).

$$Mo_2C = -3252.96 + (6.86 \cdot \text{Temp}) + (8.57 \cdot \text{Methane}) - (0.00997 \cdot \text{Temp} \cdot \text{Methane}) - \left(0.00346 \cdot \text{Temp}^2\right) \quad (5)$$

In Fig. 4, the residuals follow the line closely, with no obvious outliers and no defined pattern that would suggest a transform of the analysis was required for a better fit. The Box-Cox plot did not suggest a power transform. The Cook's distance plot provides a means of determining any potential model points that do not fit the model, and is a measure of how much the model would change, should that point be deleted. A "large" value, indicated by the red line at the top of the plot, would require further investigation. All points in Fig. 4 are well below the line, indicating all are valid model points that do not, individually, overly influence the model [15].

Model Confirmation

Confirmatory experiments were performed to evaluate the model's ability to predict Mo_2C synthesis. Three experiments were conducted within the design space—869 °C, 70% CH_4, and a 2 h hold—at parameters that would maximize Mo_2C synthesis, while keeping other factors within the range of the design space. The predicted mean for the selected parameters was 97.11% Mo_2C, with a 95% confidence interval of 90.5%—103.1% Mo_2C. The average of the three confirmatory experiments was 93.23% Mo_2C; the model confirmation was successful within the 95% confidence interval.

Conclusion

Molybdenum carbide synthesis was successfully modeled in a fluidized-bed reactor using Design-Expert12® analysis software. Three factors were evaluated over seventeen experiments to determine their effects on Mo_2C synthesis: reaction temperature, time, and inlet gas composition. Of the three variables, reaction temperature and gas composition were found to be statistically significant. Time was not a significant variable, in the analyzed range of 2.0 to 6.0 h. The developed model was found to have an R^2 of 0.9180, and an equation in terms of actual factors was developed. Three confirmatory experiments were performed, which validated the model. The synthesis of molybdenum carbide at greater than 99% conversion efficiency was achieved, at temperatures less than 900 °C.

Acknowledgements Research was sponsored by the Combat Capabilities Development Command Army Research Laboratory and was accomplished under Cooperative Agreement Number W911NF-20-2-0163. The views and conclusions contained in this document are those of the authors and should not be interpreted as representing the official policies, either expressed or implied, of the Combat Capabilities Development Command Army Research Laboratory or the U.S. Government. The U.S. Government is authorized to reproduce and distribute reprints for Government purposes notwithstanding any copyright notation herein.

References

1. Guil-Lopez R, Nieto E, Botas JA, Fierro JLG (2012) On the genesis of molybdenum carbide phases during reduction-carburization reactions. J Solid State Chem 190:285–295
2. Khabbaz S, Honarbakhsh-Raouf A, Ataie A, Saghafi M (2013) Effect of processing parameters on the mechanochemical synthesis of nanocrystalline molybdenum carbide. Int J Refract Metals Hard Mater 41:402–407
3. Chorney MP, Downey JP, Sudhakar KV (2022) Evaluation of processing parameters for the production of tungsten carbide in a fluidized bed reactor. In: 10th international symposium on high-temperature metallurgical processing. Springer, Cham, pp 393–401
4. Lee JS, Oyama ST, Boudart M (1987) Molybdenum carbide catalysts. J Catal 125–133:106
5. Mehdad A, Jentoft RE, Jentoft FC (2019) Single-phase mixed molybdenum-tungsten carbides: synthesis, characterization and catalytic activity for toluene conversion. Catal Today 323:112–122
6. Wang X-H, Hao H-L, Zhang M-H, Li W, Tao K-Y (2006) Synthesis and characterization of molybdenum carbides using propane as carbon source. J Solid State Chem 179:538–543
7. Vitale G, Guzman H, Frauwallner ML, Scott CE (2015) Synthesis of nanocrystalline molybdenum carbide materials and their characterization. Catal Today 250:123–133
8. Wang H, Wang Z, Chen S (2012) Preparation of molybdenum carbides with multiple morphologies using surfactants as carbon sources. J Solid State Chem 194:19–22
9. Nayak SK, Benavldez AD, Matanovic I, Garzon FH (2019) Thermochemical analysis of Mo–C–H system for synthesis of molybdenum carbides. Thermochim Acta 676:27–32
10. Wallace GC, Downey JP, Chorney JL, Schumacher K, Mallard A (2018) Synthesis of carbide ceramics via reduction of adsorbed anions on an activated carbon matrix. In: 8th international symposium on high-temperature metallurgical processing. Springer, Cham, pp 125–134
11. Cheremisinoff NP, Cheremisinoff PN (1984) Hydrodynamics in fluidization. In: Hydrodynamics of gas-solids fluidization. Gulf Publishing Company, Houston, pp 137–206

12. Kunii D, Levenspiel O (1991) Fluidization engineering, 2nd edn. Butterworth-Heinemann, Oxford
13. Weimer AW (1997) Fluidized bed reactor processes. In: Carbide, nitride, and boride materials synthesis and processing. Chapman & Hall, London, pp 169–180
14. Cullity BD (1978) In: Elements of X-ray diffraction, 2nd edn. Addison-Wesley Publishing Company, Inc., Reading, pp 383–396
15. Stat-Ease, "Cook's Distance," Stat-Ease, Inc.; Design-Expert, 2022. [Online]. Available: https://www.statease.com/docs/v12/screen-tips/analysis-node/diagnostics/cooks-distance/. Accessed August 2022

Fabrication of Ultra-Lightweight and Highly Porous Alumina Scaffolds by a Novel Sol–Gel/Freeze Casting Hybrid Method

Pei-Chieh Ho, Haw-Kai Chang, and Po-Yu Chen

Abstract Ceramic-based scaffolds developed by the freeze casting method exhibit anisotropic lamellar and interconnected porous structure and can be adopted for filtration, insulation, absorption, and many applications. However, the upper limit of porosity caused by the unstable mechanical properties from low solid-loading slurries has restricted functionalities of scaffolds fabricated by the traditional freeze casting method. In this study, the sol–gel/freeze casting hybrid method was developed to fabricate the alumina scaffolds with low bulk density (200–500 kg/m^3) and proper specific strength. The microstructural features of the lamellar structure and continuous surface developed from the condensation reaction were evaluated by SEM. The ultra-lightweight porous alumina scaffolds successfully fabricated by this hybrid method show high specific surface area and proper mechanical stability. The porosity of alumina scaffolds can reach over 90%, possessing great potential for filtration and gas absorption applications in the future, and this hybrid sol–gel/freeze casting approach can be extended to ceramic/glass scaffolds with varying functionalities.

Keywords Porous ceramics · Ultra-lightweight · Freeze casting · Sol–gel method · Alumina

Introduction

In the last few decades, lightweight materials have been developed with a wide diversity and applied to various fields. Many optimized structures in nature provide the inspirations of novel design in lightweight structure, such as woods [1], human bone [2, 3], and fruit peels [4, 5]. Moreover, among artificial materials, aerogel is regarded as a representative of ultra-lightweight ceramics which has ultra-high specific surface area (500–1200 m^2/g) and porosity (80–99.8%), which results in extremely low density (~0.003 g/cm^3) [6]. Porous ceramic can effectively reduce the weight of materials with the existence of open and closed pores [7]. Through varying synthesis

P.-C. Ho · H.-K. Chang · P.-Y. Chen (✉)
National Tsing Hua University, Hsinchu City, Taiwan
e-mail: poyuchen@mx.nthu.edu.tw

B. Li et al. (eds.), *Advances in Powder and Ceramic Materials Science 2023*, The Minerals, Metals & Materials Series, https://doi.org/10.1007/978-3-031-22622-9_4

strategies, porous ceramics with a wide range of porosity and wide pore size distribution can be fabricated by the sol–gel method [8, 9], sacrificial template [10], partial sintering [11], direct foaming [11], and additive manufacturing [12]. Freeze casting is one of the sacrificial template methods using ice crystal as templates to fabricate anisotropic porous scaffolds and has the characteristic of cost-effective advantage, wide material selectivity and controllable process. Typical freeze casting technique proceeds in four steps: slurry preparation, solidification, sublimation, and sintering. Directional freezing gives rise to the anisotropic growth of ice crystals repelling the ceramic particles in slurry, thus generating significant microscale lamellar structures [13–15].

However, in the conventional freeze casting technique, the pore sizes are limited in micrometer scale owing to the growing velocity of ice dendrites. Besides, the scaffolds fabricated by freeze casting method have the upper limit on porosity since the scaffolds with lower solid loading would cause unacceptable loss in mechanical properties [14, 15]. Moreover, collapsing and cracks formation in the scaffolds may occur after freeze drying or sintering. To address these limitations, the sol–gel method was integrated with conventional freeze casting system to synthesize ultra-lightweight scaffolds with anisotropic lamellar structure [16].

The sol–gel method is a wet-chemical technique developed for a long time to prepare ceramics and nanomaterials in diverse forms, such as bulks, powders, films, and fibers, which can be broadly applied to various fields [9]. The sol–gel process is the reaction in which precursors undergo the hydrolysis and condensation process, and it ends up in two stages, sol and gel, respectively [17]. Metal alkoxide is the most common precursor used in sol–gel process and has been extensively studied to date, which reacts with water easily while it is immiscible with water; therefore, a common solvent, usually the alcohol, has to be added in the solution [18]. The primary advantages of sol–gel processing are the low reaction temperature, homogeneity of product, and a wide variety in material selection.

Typical freeze casting system was combined with sol–gel method to obtain the hierarchical ultra-lightweight scaffolds with anisotropic lamellar structure, which consists of microscale and nanoscale pores formed by the sublimation of ice templates and pyrolysis of organic additives [16]. This study aims at utilizing the sol–gel/freeze casting hybrid method to fabricate highly porous alumina scaffolds successfully. Since alumina is a popular low-cost material widely used as a catalyst and wear-resistant material with a great strength, heat resistance, and biocompatibility [19], it is chosen as the main material in this research. The porosity of alumina scaffolds synthesized from this hybrid method can reach over 90%, which is higher than that of most ceramic-based scaffolds fabricated from conventional freeze casting technique [14], possessing great potential for filtration and gas absorption applications. On top of that, this hybrid sol–gel/freeze casting approach can be extended to other ceramic/glass scaffolds with varying functionalities.

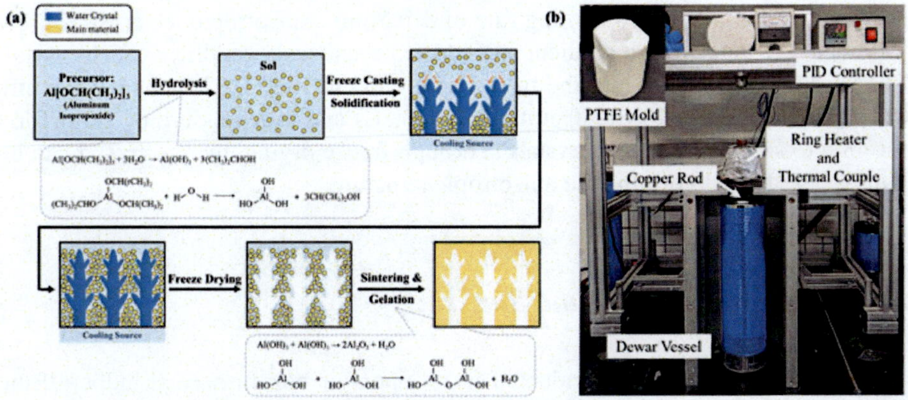

Fig. 1 **a** Flow chart of the sol–gel/freeze casting hybrid method to fabricate alumina scaffolds **b** Photo shows the freeze casting system and its main elements: a Dewar vessel, a Teflon mold, a copper stick, a thermal couple, a heating ring, and a PID controller

Experimental Method

Slurry Preparation for Sol–Gel/Freeze Casting Hybrid Method

Figure 1A represents the flow chart for the preparation of alumina scaffold. AIP (Aluminum isopropoxide, ≥98%, Sigma Aldrich) was used as the precursor in sol–gel method to prepare the slurry based on the following steps. Firstly, AIP was mixed with the isopropyl alcohol (i-PrOH), distilled water (H_2O), and hydrogen chloride (HCl). Then, the mixture was placed in the closure system and subsequently stirred at 70 °C for 12 h to promote hydrolysis of AIP and form milky sol. The molar ratio of AIP: i-PrOH: H_2O: HCl was fixed at 1: 30: 60: 0.11. Secondly, the sol was heated in an oven at 60 °C for 24 h to remove hydrogen chloride and excessive water and get the yellowish transparent product, amorphous aluminum hydroxide, from AIP hydrolysis. The slurry was prepared by dissolving aluminum hydroxide in distilled water dissolving 0.5wt% polyethylene glycol (PEG, MW = 6000, Alfa Aesar) as organic binder. The concentration of the slurry was controlled at 5 and 10vol%, which affects the porosity of scaffolds.

Freezing Process and Removal of Ice

The freeze casting system is composed of a Dewar vessel, a cylindrical copper stick, a thermal couple, a Teflon mold, and a PID controller, as shown in Fig. 1b. Liquid nitrogen was poured into the Dewar vessel as the cooling source to solidify the slurry. The PID connected with the copper stick was utilized to stably control the cooling rate of slurry. The slurries with different concentrations were poured into the Teflon mold and solidified at controlled cooling rate of 5 °C/min. Anisotropic ice crystals grew following the temperature gradient and formed microscale dendrites of ice crystal. The specimens were put into the freeze dryer (FD 5030/8530, Panchum Scientific Corp., Taiwan) under low temperature (<−80 °C) and low pressure (<200 mTorr) for 48 h to sublimate the ice crystals. Through freeze drying, the ice crystal can be removed without deforming the anisotropic structure.

Condensation Reaction and Sintering

The sintering process was conducted in an open-air box furnace (Lindberg/Blue MTM BF51314C, Thermal Product Solution, USA). The green bodies of scaffolds were heated up from room temperature to 1400 °C with a constant heating rate of 4 °C/min, held for 4 h, and then cooled to room temperature. There are three purposes of sintering. First, the condensation reaction catalyzed by high temperature will cause dehydration and aluminum hydroxide molecules cross-linking to amorphous alumina, finishing whole sol–gel method of alumina synthetization. Second, the sintering process is applied to improve the mechanical stability and structural completeness in scaffolds; moreover, maintaining temperature at 1400 °C can facilitate the phase transformation from amorphous alumina to α-alumina. Last, the organic binder will be burned out in the heating process.

Characterization

Phase Analysis

The alumina scaffolds were characterized by SHIMADZU X-ray diffractometer (HT-XRD-6000, Shimadzu Co., Japan) with Cu $k\alpha(\lambda = 1.54\,\text{Å})$ radiation. Phase identification based on the intensities radiated from specimens was carried out utilizing MDI Jade 6.0 program in comparison with JCPDF standard cards.

Observation of Microstructure

The microstructural features of scaffolds in transverse and longitudinal planes were observed by a scanning electron microscopy (Hitachi SU-8010, Hitachi High Technologies America, Inc., USA) after being platinum-coated. The accelerating voltage was 15 keV, and the working distance was 8 mm.

Determination of Porosity

Matsuhaku Density Tester (TES-214 k, Group Prospers Enterprise Co., Taiwan) was used to measure the porosity of alumina scaffolds. Porosity is defined as the fraction of void volume divided by the bulk volume in materials. The measurements were based on Archimedes' principle. By filling liquid into the pores in scaffolds, the experimental porosity can be calculated by the following equation:

$$\text{experimental porosity}(\%) = \frac{(W_2 - W_1) \times \rho_l}{W_3 - W_2 / \rho_1}$$

W_1(g) is the mass of scaffolds in the air, W_2(g) is the mass of scaffolds filled with liquid in the air, and W_3(g) is the mass of scaffolds under liquid. $\rho_l$represents the liquid density (g/m^3). Absolute ethanol (Sigma–Aldrich, St. Louis, USA) was used to fill the scaffold in the measurement, so the theoretical density was set to 0.79 g/cm^3.

Compression Test

A universal mechanical testing system (Instron 3343 Single Column Testing System, Norwood, MA, USA) equipped with a 1 kN load cell was used to measure the compressive mechanical properties of scaffolds. Cylindrical scaffolds were 9.52 $\pm$ 0.16 mm in diameter and 13.93 $\pm$ 0.27 mm in height. The compression test was conducted with constant strain rate of 0.001 s^{-1}. The load (N) and displacement (mm) within measurement were recorded to calculate the compressive strength and Young's modulus.

Results and Discussion

Material Confirmation and Phase Analysis

After 4 h sintering, the scaffold weight yields approximately 66.7 $\pm$ 3.3%, which means it surely condensed to alumina because the weight loss in sintering process matches with that of the condensation reaction of $Al(OH)_3$. According to [20], it is also a simple method to identify whether the form of aluminum hydroxide is $Al(OH)_3$ after the hydrolysis of aluminum alkoxide that depends largely on the reaction conditions.

Figure 2a shows the XRD pattern of alumina scaffolds after scaffolds were sintered at 1000 and 1400 °C, respectively. The phase of alumina scaffolds sintered at 1000 °C is obviously amorphous and shows poor crystallinity. Nevertheless, some characteristic peaks radiated closely match with the JCPDS standard card of α-alumina with

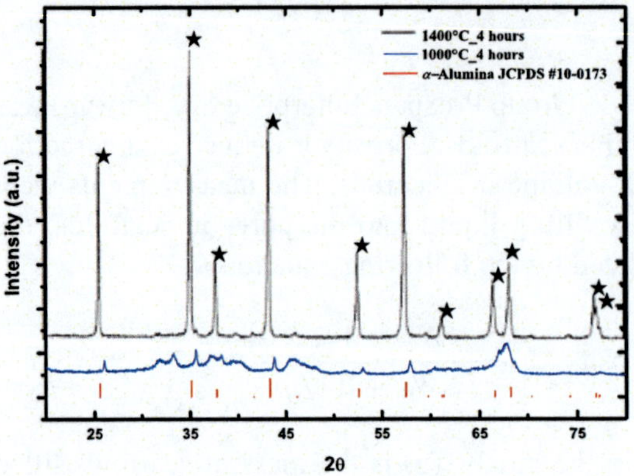

Fig. 2 **a** XRD pattern of alumina scaffolds sintering at 1000 and 1400 °C for 4 h **b** Photograph of the ultra-lightweight alumina scaffold

slight shift. For scaffolds sintering at 1400 °C, its XRD pattern can be significantly indexed with α-alumina (JCPDS 10–0173). With complete coordination between characteristic peaks and standard cards, no processing residue or secondary phases are found in the alumina scaffolds.

Extrinsic Appearance and Microstructure Features of Scaffolds

As shown in Fig. 2b, the ultra-lightweight alumina scaffolds can be successfully fabricated and maintain the structural stability to such a degree without significant crack formation and irregular shrinkage after sintering process.

SEM micrographs of transverse sections of alumina scaffolds with the slurry solid contents of 5 and 10vol % are shown in the Fig. 3a, d, respectively. It can be observed that scaffolds with 10vol% solid content have obvious lamellar structure on transverse section. However, for 5vol% solid content, because less particles existed in slurry, the ice dendrite suffered from relatively small resistance and tend to form cellular structure instead of lamellar structure. The distances between neighboring lamellae are approximately 56.15 ± 12.05 μm and 35.16 ± 6.19 μm, and the corresponding solid contents are 5 and 10vol%. Figure 3b, e are SEM micrographs of longitudinal sections synthesized at 5 and 10vol% solid contents correspondingly, both of which reveal the stacking layers of lamellae caused by the unidirectional ice growth.

In comparison with the typical freeze casting method, Fig. 3c shows that the sol–gel derived lamellae in scaffolds are quite smooth. The nanometer scale pores on the surface of lamellae are resulted from the burnout of organic binder in slurry [16]. With the decreasing of solid content, the nanoscale pore size becomes larger and smooth surface turns out to be discontinuous in Fig. 3f. Despite the discontinuous surface

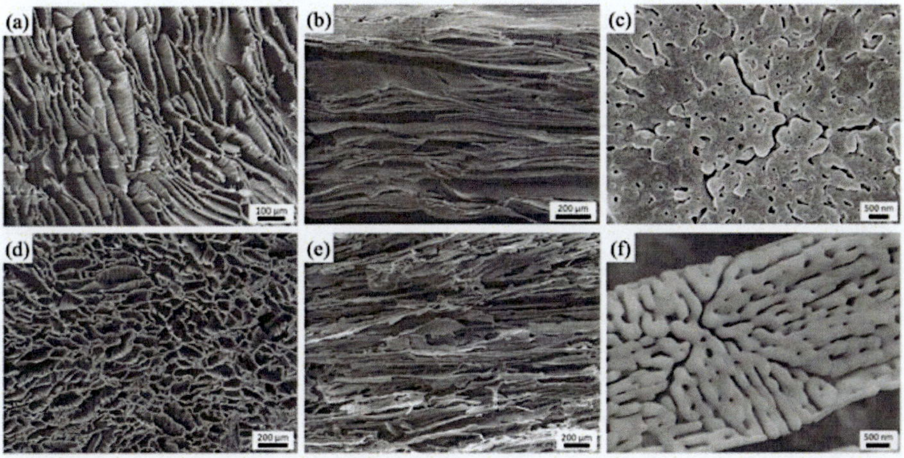

Fig. 3 SEM micrographs of alumina scaffolds in **a** in transverse section with 10vol% solid content **b** in longitudinal plane with 10vol% solid content **c** smooth lamellae surface of alumina scaffolds with nanoscale pores with 10vol% solid content **d** transverse section with 5vol% solid content **e** longitudinal plane with 5vol% solid content **f** discontinuous lamellae surface of alumina scaffolds with 10vol% solid content

of lamellae in scaffolds, hierarchical porous structure is formed by sol–gel/freeze casting hybrid method in ultra-lightweight alumina scaffolds indeed.

Porosity, Pore Area, and Pore Size Distribution

Typical freeze casting method has the difficulties fabricating highly porous scaffolds through the extremely low solid content in slurry because of the unacceptable loss in mechanical properties, uneven shrinkage, crack formation, and even collapsing. The porosity of literature scaffolds fabricated by conventional freeze casting is plotted in Fig. 4 [21–32]. The porosity of alumina scaffolds synthesized from conventional freeze casting ranges between 40 and 65%, and those of other ceramic-based scaffolds range from 15 to 87%. In comparison with typical freeze casting, porosity of scaffolds fabricated by the sol–gel/freeze casting hybrid process has the lower limit at approximately 85% because the slurry would become viscous with higher concentration and lose fluidity, which is caused by high concentration of nanoscale particles in sol, also regarded as colloid solution. Consequently, through sol–gel/freeze casting hybrid method, scaffolds can be equipped with higher porosity than scaffolds fabricated from conventional freeze casting process and also possess relatively low density, as shown in Table 1.

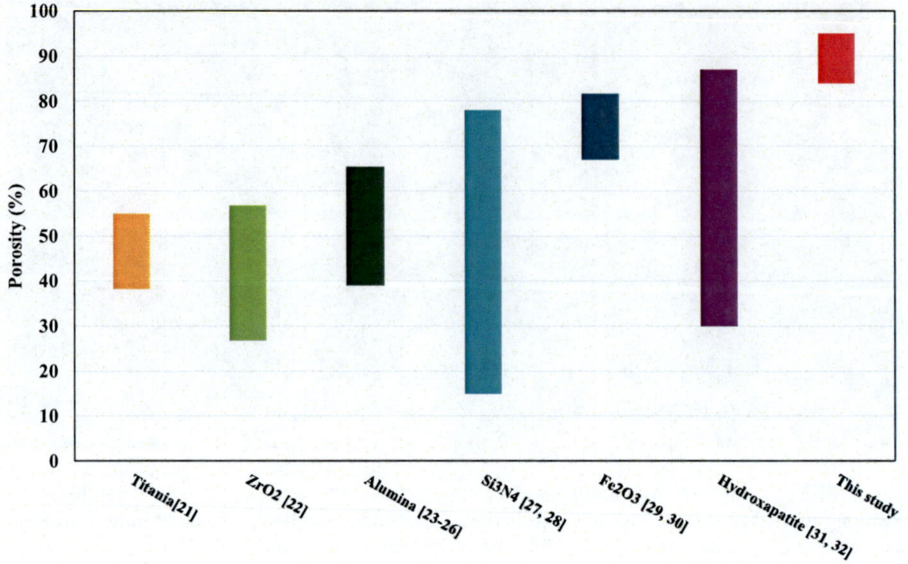

Fig. 4 Comparison between porosity of ceramic-based scaffolds fabricated by conventional freeze casting and by that of the scaffolds synthesized by sol–gel/freeze casting hybrid method in this study

Table 1 General characteristics and compressive mechanical properties of scaffolds with 5vol% and 10vol% solid content synthesized by sol–gel/freeze casting hybrid method

Parameters	Porosity (%)	Density (g/cm^3)	Relative density	Young's modulus (kPa)	Compressive strength (MPa)	Specific compressive strength (MPa)
5vol%, 1400 °C	93.7 ± 1.2	0.21 ± 0.02	0.05 ± 0.01	12.35	5.47	26.05
10vol%, 1400 °C	88.8 ± 2.1	0.27 ± 0.04	0.07 ± 0.01	35.07	10.47	38.78

Compressive Mechanical Properties

Compressive mechanical properties synthesized from sol–gel/freeze casting hybrid method are revealed in Table 1. The cylindrical scaffolds have the average of 9.52 ± 0.16 mm in diameter and 13.93 ± 0.27 mm in height. The densities of scaffolds synthesized by 5vol% and 10vol% aluminum hydroxide slurry are 0.21 ± 0.0.2 g/cm^3 and 0.27 ± 0.04 g/cm^3, respectively. The results show that average Young's modulus of 10vol% solid content is roughly 2.84 times higher than that of 5vol% solid content (35.07 kPa). Besides, the collapsing process of scaffolds is shown in Fig. 5. The cylindrical scaffolds would break gradually layers by layers

Fig. 5 Photograph of the collapsing process during compressive test of alumina scaffolds with 10vol% solid content

owing to the interior unidirectional structure. The scaffolds with vertical wall alignments also have better compressive mechanical properties in the direction parallel to the solidification direction [33].

Conclusions

Sol–gel/freeze casting hybrid method has been established in this study to successfully fabricate the ultra-lightweight and highly porous alumina scaffolds. In the experimental step, the slurry prepared by the hydrolysis of precursor can be easily solidified and maintains the stable structure in both exterior appearance and interior microstructure of alumina scaffolds. Besides, the lamellar structure and cellular structure are obvious in the transverse plane of scaffolds with 10 and 5vol% solid content correspondingly. Nanoscale pores form on lamellae surfaces due to the organic binder burnout. The porosity of scaffolds depends on the solid content of slurry, ranging between 85 and 95%, which is higher than scaffolds fabricated by typical freeze casting. Moreover, alumina scaffolds synthesized from this hybrid method have the proper compressive strength and structural integrity. This hybrid method has advantage to be extended to other ceramics-based scaffolds with diverse material selection and fabricate ultra-lightweight and highly porous scaffolds, and may further lead to various potential applications.

References

1. Gibson LJ (2012) The hierarchical structure and mechanics of plant materials. J R Soc Interface 9(76):2749–2766
2. Müller R (2009) Hierarchical microimaging of bone structure and function. Nat Rev Rheumatol 5(7):373–381

3. Reznikov N, Shahar R, Weiner S (2014) Bone hierarchical structure in three dimensions. Acta Biomater 10(9):3815–3826
4. Zhang W, Yin S, Yu TX, Xu J (2019) Crushing resistance and energy absorption of pomelo peel inspired hierarchical honeycomb. Int J Impact Eng 125:163–172
5. Foo KY, Hameed BH (2012) Porous structure and adsorptive properties of pineapple peel based activated carbons prepared via microwave assisted KOH and K2CO$_3$ activation. Microporous Mesoporous Mater 148(1):191–195
6. Dorcheh AS, Abbasi MH (2008) Silica aerogel; synthesis, properties and characterization. J Mater Process Technol 199(1–3):10–26
7. Zdravkov B, Čermák J, Šefara M, Janků J (2007) Pore classification in the characterization of porous materials: a perspective. Open Chem 5(2):385–395
8. Cividanes LS, Campos T, Rodrigues LA, Brunelli DD, Thim GP (2010) Review of mullite synthesis routes by sol–gel method. J Sol-Gel Sci Technol 55(1):111–125
9. Guo X, Zhang Q, Ding X, Shen Q, Wu C, Zhang L, Yang H (2016) Synthesis and application of several sol–gel-derived materials via sol–gel process combining with other technologies: a review. J Sol-Gel Sci Technol 79(2):328–358
10. Yin L, Zhou X, Yu J, Wang H (2016) Preparation of silicon nitride foam with three-dimensional interconnected pore structure. Mater Des 89:620–625
11. Ohji T, Fukushima M (2012) Macro-porous ceramics: processing and properties. Int Mater Rev 57(2):115–131
12. Zhao J, Zhang M, Zhu Y, Li X, Wang L, Hu J (2019) A novel optimization design method of additive manufacturing oriented porous structures and experimental validation. Mater Des 163:107550
13. Deville S (2008) Freeze-casting of porous ceramics: a review of current achievements and issues. Adv Eng Mater 10(3):155–169
14. Deville S (2010) Freeze-casting of porous biomaterials: structure, properties and opportunities. Materials 3(3):1913–1927
15. Scotti KL, Dunand DC (2018) Freeze casting–a review of processing, microstructure and properties via the open data repository, FreezeCasting. net. Progress Mater Sci 94:243–305
16. Chang HK, Chen PY (2020) Synthesis of silica-based scaffolds with high porosity and controllable microstructure by a sintering-free sol–gel/freeze-casting hybrid method under mild conditions. J Market Res 9(6):16167–16178
17. Zhang B (2018) Principles, methods, formation mechanisms, and structures of nanomaterials prepared in the liquid phase. In: Zhang B (ed) Physical fundamentals of nanomaterials. William Andrew Publishing, Boston, pp 71–111
18. Livage J (2011) Inorganic materials, sol-gel synthesis of. Encycl Mater: Sci Technol 4105–4107
19. Figiel P, Rozmus M, Smuk B (2011) Properties of alumina ceramics obtained by conventional and non-conventional methods for sintering ceramics. J Achiev Mater Manuf Eng 48(1):29–34
20. Yoldas BE (1973) Hydrolysis of aluminium alkoxides and bayerite conversion. J Appl Chem Biotech 23(11):803–809
21. Li JC, Dunand DC (2011) Mechanical properties of directionally freeze-cast titanium foams. Acta Mater 59(1):146–158
22. Naleway SE, Fickas KC, Maker YN, Meyers MA, McKittrick J (2016) Reproducibility of ZrO_2-based freeze casting for biomaterials. Mater Sci Eng, C 61:105–112
23. Hautcoeur D, Gonon M, Baudin C, Lardot V, Leriche A, Cambier F (2018) Alumina porous ceramics obtained by freeze casting: structure and mechanical behaviour under compression. Ceramics 1(1):83–97
24. Souza DF, Nunes EH, Pimenta DS, Vasconcelos DC, Nascimento JF, Grava W, Vasconcelos WL (2014) Synthesis and structural evaluation of freeze-cast porous alumina. Mater Charact 96:183–195
25. Miller SM, Xiao X, Faber KT (2015) Freeze-cast alumina pore networks: effects of freezing conditions and dispersion medium. J Eur Ceram Soc 35(13):3595–3605
26. Sofie SW, Dogan F (2001) Freeze casting of aqueous alumina slurries with glycerol. J Am Ceram Soc 84(7):1459–1464

27. Hu HL, Zeng YP, Xia YF, Yao DX, Zuo KH (2014) High-strength porous Si3N4 ceramics prepared by freeze casting and silicon powder nitridation process. Mater Lett 133:285–288
28. Xia Y, Zeng YP, Jiang D (2012) Microstructure and mechanical properties of porous Si3N4 ceramics prepared by freeze-casting. Mater Des 33:98–103
29. Wilke SK, Mack JB, Kenel C, Dunand DC (2021) Evolution of directionally freeze-cast Fe_2O_3 and Fe_2O^{3+} NiO green bodies during reduction and sintering to create lamellar Fe and Fe-20Ni foams. J Alloy Compd 889:161707
30. Sepúlveda R, Plunk AA, Dunand DC (2015) Microstructure of Fe2O3 scaffolds created by freeze-casting and sintering. Mater Lett 142:56–59
31. Fu Q, Rahaman MN, Dogan F, Bal BS (2008) Freeze casting of porous hydroxyapatite scaffolds. I. Processing and general microstructure. J Biomed Mater Res Part B: Appl Biomater: Off J Soc Biomater, Jpn Soc Biomater, Aust Soc Biomater Korean Soc Biomater 86(1):125–135
32. Deville S, Saiz E, Tomsia AP (2006) Freeze casting of hydroxyapatite scaffolds for bone tissue engineering. Biomaterials 27(32):5480–5489
33. Niksiar P, Su FY, Frank MB, Ogden TA, Naleway SE, Meyers MA, Porter MM (2019) External field assisted freeze casting. Ceramics 2(1):208–234

Effect of Three-Dimensionally Connected Porous Hydroxyapatite Ceramics on Enhancing Heat Storage of Lithium Nitrate Phase Transformation Materials

Ruifan Zhou, Shuang Song, Jinhong Li, Lu Jiang, and Yixiu Xin

Abstract As a medium-temperature phase change energy storage material, lithium nitrate has many applications in phase change energy storage due to its excellent thermal properties and specific heat capacity. However, its inherent leakage and corrosion problems have adversely affected its continued development. A unique pore structure must be designed to solve the severe leakage problem. The porous hydroxyapatite (Hap) ceramic (PHC) prepared by the microemulsion template method has a pore-window structure, uniform and controllable pore size, and three-dimensional permeability. Porous ceramics with different porosity can be obtained by adjusting the solid content, based on 80.2% porous ceramic. The encapsulation of $LiNO_3$ into permeable ceramic channels by the medium-temperature melting method can solve PCM's corrosion and leakage problems. In this work, the phase transition temperatures during melting and solidification are 255.7 and 235.8 °C, and the latent heats are 240.2 and 252.2 $J{\cdot}g^{-1}$, respectively. The sample solids content is 35 wt%, and the maximum packing ratio is 78 wt%. Therefore, the prepared composite phase change materials (CPCMs) possess controllable three-dimensional pore structures, excellent chemical properties, cycling stability, and chemical stability.

Keywords Hydroxyapatite · Three-dimensional connected porous · Microemulsion template method · Lithium nitrate

---

The authors contributed equally to this work.

R. Zhou · S. Song · J. Li (✉) · L. Jiang · Y. Xin
Beijing Key Laboratory of Materials Utilization of Nonmetallic Minerals and Solid Wastes, National Laboratory of Mineral Materials, School of Materials Science and Technology, China University of Geosciences, Beijing 100083, PR China
e-mail: jinhong@cugb.edu.cn

B. Li et al. (eds.), *Advances in Powder and Ceramic Materials Science 2023*, The Minerals, Metals & Materials Series, https://doi.org/10.1007/978-3-031-22622-9_5

Introduction

Energy and environmental issues have always been the focus of social development. Reducing energy consumption and saving energy have always been the focus of current research [1]. Solid–liquid PCMs had the advantages of compact structure, small volume change, and high heat storage density [2]. They are usually used in building energy, intelligent temperature control textile, microelectronic thermal management, solar energy, residual waste heat recovery, and other fields [3]. Medium temperature inorganic PCMs, such as nitrate and carbonate, are widely used in solar collectors with appropriate PCMs [4], good thermal stability, low undercooling, and significant phase change latent heat. In particular, nitrate PCMs have the advantages of low cost, low corrosion, and high strength [5]. Compared with other medium and high-temperature molten salts, nitrates have a higher specific heat capacity and enthalpy values and are considered to have greater application prospects. The excellent properties of lithium nitrate have attracted attention in phase change materials [6].

However, leakage problems limit the application of nitrates. It is necessary to encapsulate them in porous media to solve the leakage problem [7]. The porous ceramics sintered by the microemulsion method do not contain other contaminants. At the same time, the uniformity of the pore structure can be ensured so that the phase change material can be uniformly attached to the entire carrier rather than aggregated in one place, significantly reducing the overall heat storage performance of the material [8].

Hydroxyapatite (Hap) is a significant component of human and animal bones [9]. It is widely used as an adsorbent material. It is non-toxic, harmless, inexpensive, easy to obtain, has good compatibility, and strong adsorption [10, 11]. There are relatively few reports on Hap as a multi-empty packaging material. Consider the preparation of Hap porous ceramics by the microemulsion method. Compared with the traditional porous mineral carrier, it has the advantages of uniform pore distribution, controllable pore size, and no impurities [12]. At the same time, the unique benefits of Hap provide a new direction for composite materials in biological applications.

This study used the microemulsion template method to prepare porous hydroxyapatite ceramics (PHC) with controllable pore structures. The solid content was controlled to be 35–55 wt%. A novel structure stable composite PCMs (CPCMs) was prepared by impregnating $LiNO_3$ (PHCL) at high temperatures. The small window improves its adsorption capacity. At the same time, adjusting the solid content could regulate the strength, porosity, and adsorption capacity of PHC. The results show that the three-dimensional pore-window structure can effectively improve the adsorption capacity and heat storage capacity of PCMs. PHCL was a CPCMs with stable morphology, excellent performance, high enthalpy, and good chemical stability.

Experimental

Materials

Commercial Hap powder with a particle size of 60 nm was used (A. R., main impulse content was less than 0.35%). Polyacrylate (Adamas Company) was used as a dispersant to prevent particle aggregation. N-propyl gallate (Sinopharm Chemical Reagent Co., Ltd.) was used as an emulsifier, and ammonia (Sinopharm Chemical Reagent Co., Ltd.) adjusted pH. The oil phase of the second stage of the template method uses n-octane with a volume fraction of 200% (Beijing Chemical Plant) and phase change material $LiNO_3$ (Beijing Chemical Reagent Co., Ltd.).

Preparation of PHC

Porous Hap foam ceramics were prepared by the microemulsion template method, as shown in Fig. 1a. Firstly, polyacrylate is added to a certain amount of deionized water as the dispersant, and then, it is shaken and dispersed. Then, the commercially available was prepared from 35 wt% to 55 wt%, ammonia was adjusted to pH 10, and the precursor was added to the ball milling tank for 20 h to obtain a uniformly dispersed Hap suspension. The propyl gallate and ethanol were disproportionately dissolved, added to the rest for 30 min as an emulsifier, and added to the break with octane. The ratio of oil and mixture was 1:1.5. The water mixture was formed by high-speed shearing, emulsifying the mixed liquid through a domestic agitator to create an oil-in-water emulsion. The emulsion was successfully dried for 7 days in a diameter of 90 mm until the emulsion was completely dried. The samples were sintered at 1300 °C for 2 h, heated at 5 °C/min, and the porous Hap foam ceramics were prepared by thermal insulation for 2 h. Most of the obtained samples are named PHC1-5.

Preparation of Composite PCMs

As in Fig. 1b, prepare stable composite PHCL, put PHC with different solid content (35–55 wt%) into the crucible, add $LiNO_3$ into the crucible, and fully cover the PHC. After heating to 350 °C in a muffle furnace and holding for 2 h, molten $LiNO_3$ was dissolved at medium temperature and impregnated into the unique structure of PHC. The pore-window window structure in PHC adsorbs $LiNO_3$ in the porous structure through surface tension and capillary force, and the obtained sample was named PHCL1-5. Repeat the heat treatment process five times and heat cycle PHCL until there is no leakage trace in the crucible.

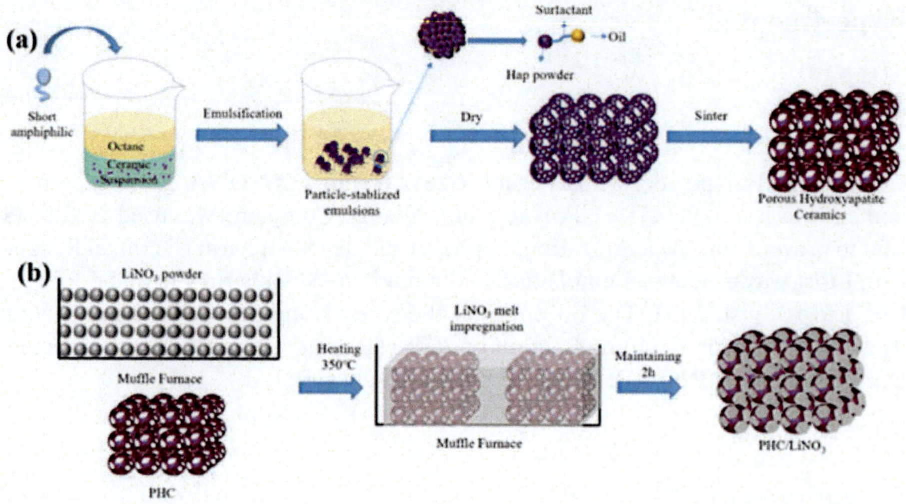

Fig. 1 Preparation process. **a** Preparation of PHC by microemulsion template method. **b** PHCL was prepared by high-temperature melt impregnation

Results and Discussion

Effects of Solid Contents on Properties of PHC

Figure 2a shows the X-ray diffraction analysis of commercial Hap powder used in the experiment and PHC with a solid content of 45 wt% sintered at 1300 °C. Data analysis showed that the prominent peaks of commercial powder correspond to the main diffraction peaks of PHC one by one and correspond to the heights of the Hap standard card (PDF#09-0432). The corresponding crystal surface is marked in Fig. 2a, which clearly shows that the phase of Hap powder has not changed significantly after high-temperature sintering. After high-temperature sintering, the diffraction peak was more substantial, indicating that dense and porous Hap ceramics can be prepared by sintering at high temperatures. Figure 2b shows the open porosity of PHC sintered at 1300°C, calculated by the Archimedes drainage method. The results are shown in Table 2. The results show that with the increase of solid content, the apparent porosity gradually decreases from 80.18% to 55.79%.

Meanwhile, it was evident from Fig. 4 that the pore diameters of PHC are distributed around 5 μm, indicating that the prepared PHC by the Microemulsion template method has a uniformly distributed pore structure, shown in Fig. 5. The spherical shape was the distribution shape of the oil phase in the Pickering emulsion, and the macroporous system was spherical from the front. The small window was distributed in the macropores without rules, which is conducive to impregnation and leakage prevention of PCM. It effectively fills a lot of PCMs, and it could effectively inhibit the leakage of PCM and was not easy to corrode. The lower solid content

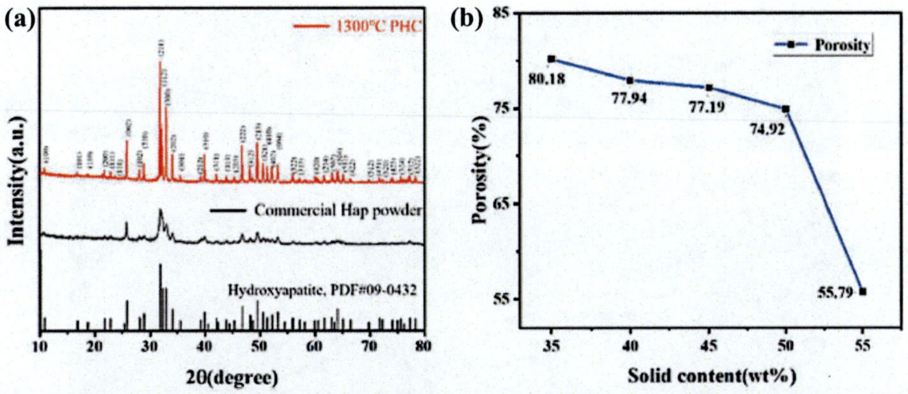

Fig. 2 **a** XRD pattern of commercial Hap powder and PHC with a solid content of 45 wt%; **b** Calculation of open porosity of PHCs with different solid content by Archimedes drainage method

of PHC (Fig. 5a) shows apparent spherical pore structure, small window structure, sparse distribution, and three-dimensional skeleton structure. With the increase of solid content from 35 wt% to 55 wt%, the stable morphology of PHC has apparent changes. The rise of Hap particles in Pickering emulsion leads to a decrease in the size of oil droplets and a significant increase in the number of PHC droplets. There was a three-dimensional skeleton structure. The accumulation of Hap particles was thickened, and the distribution of pores and windows became dense. It was not easy to distinguish between pore structure and window structure clearly. The porosity showed a downward trend, which increases the encapsulation ability of PCM to a certain extent and prevents the leakage of the PCM melting process to a certain extent (Fig. 3).

Fig. 5a FT-IR spectra and b XRD pictures of LiNO3, PHC, and PHCL

Characterization of the Form-Stable CPCMs

Form-Stabilization of the fs-CPCMs

As shown in Fig. 4, the morphology of PHCL and the distribution diagram of EDS elements are shown, which could see the leakage of $LiNO_3$ through high-temperature melting and impregnation of $LiNO_3$. The $LiNO_3$ is uniformly filled into the hole and window structure of PHC, and the small window structure inhibits the leakage of $LiNO_3$ to a certain extent.

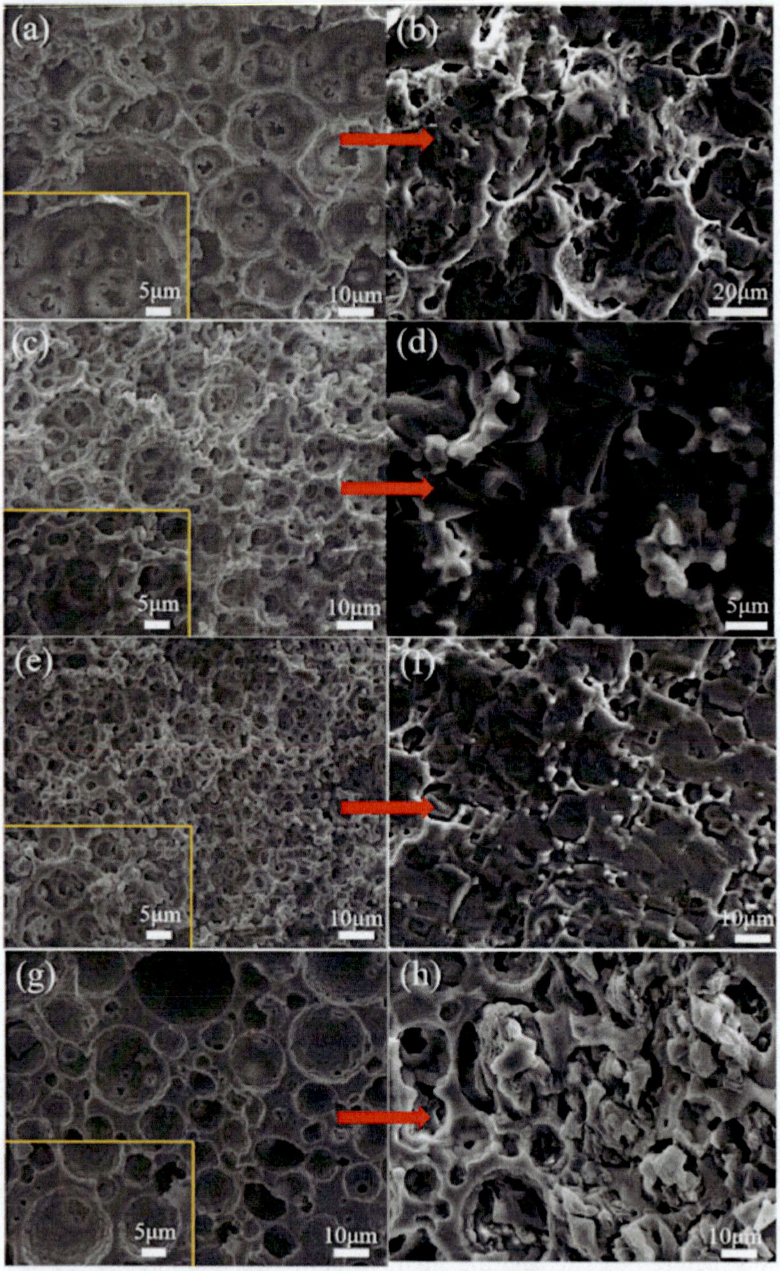

Fig. 3 The SEM pictures of PHCs sintered at 1300 °C with different solid contents and PHCL after packaging. **a** 35%, **b** 35% after packaging, **c** 45%, **d** 45% after packaging, **e** 50%, **f** 50% after packaging, **g** 55%, **h** 55% after packaging

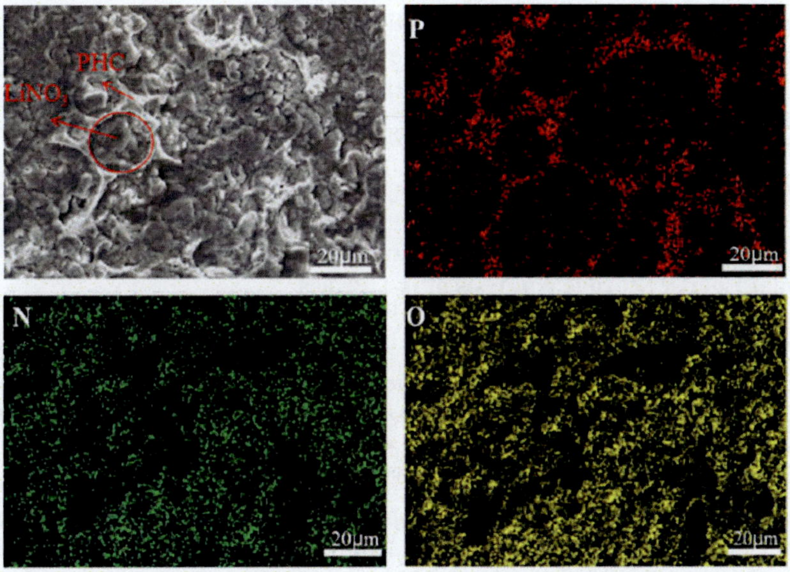

Fig. 4 EDS results of PHC after high-temperature impregnation of $LiNO_3$ (PHCL)

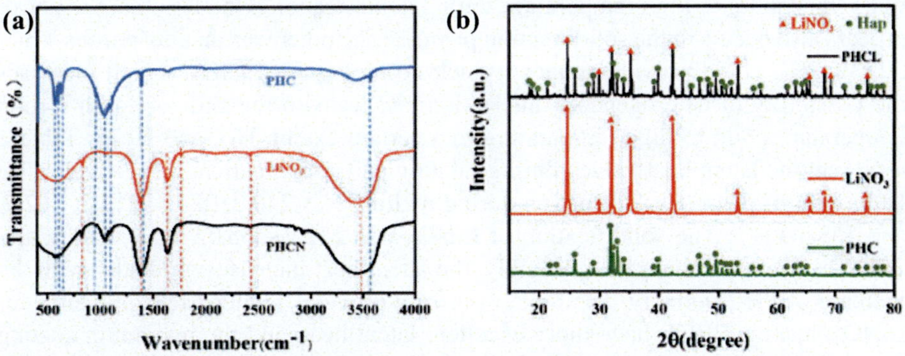

Fig. 5 **a** FT-IR spectra and **b** XRD pictures of $LiNO_3$, PHC, and PHCL

Chemical Stability Characterization of fs-CPCMs

The compatibility and chemical stability of PHC and $LiNO_3$ were tested and analyzed by XRD and FT-IR and are shown in Fig. 5a. Comparing all characteristic peaks of PHCL with those of PHC and $LiNO_3$, it could be proved that $LiNO_3$ was wrapped in PHC and did not react to produce new substances. Among them, 1090, 1043, 960, 600, and 569 cm^{-1} were the characteristic peak of apatite, 3570 and 631 cm^{-1} are the stretching vibration of hydroxyl [12], and 829, 1324, 1637, and 2425 cm^{-1} represent the antisymmetric stretching vibration region of N–O. As shown in Fig. 5b, comparing the XRD peaks of PHC, $LiNO_3$, and PHCL, it was found that the main characteristic peaks are consistent, indicating that $LiNO_3$ was successfully encapsulated into the pore structure of PHC, and they had good compatibility and crystallinity. XRD and FT-IR results show that PHC and PHCL had good stability. The impregnation process had no chemical reaction. The PHC was an excellent packaging medium for PCMs, which allows the preparation of fs-CPCMs with stable shape and chemical stability.

Thermal Energy Storage Analysis of fs-CPCMs

DSC studied the melting and solidification processes of fs-CPCMs and $LiNO_3$. As shown in Fig. 6, the PCMs and latent heat during melting and solidification are shown. It was evident from the DSC curve that the curve changes of fs-CPCMs and $LiNO_3$ were the same, and the endothermic and exothermic peaks mainly appeared at 250 °C. However, because the hydroxyl group in Hap usually exists in the form of a hydrogen bond, the phase transition temperature point of PHCLs moves slightly to a high temperature near 260 °C compared with the phase transition temperature of $LiNO_3$. During the melting process, $LiNO_3$ must overcome the intermolecular force and hydrogen bond force simultaneously, shifting the melting temperature point to a high temperature. There are two main reasons for changing phase transformation temperature to high temperature during solidification. One was that the above process also occurs in the solidification process. The other was that the porous structure of PHC could provide abundant nucleation sites for $LiNO_3$, which increases the possibility of heterogeneous nucleation and leads to the shift of solidification temperature point to high temperature to a certain extent. $H_{M.}$ and $H_{S.}$ in Table 1 represent the latent heat values during melting and solidification, respectively. The latent heat of phase transformation during melting was 240.2, 221.4, 211.2, 192.2, and 153.5 $J{\cdot}g^{-1}$. The solidification of $LiNO_3$ was 252.2, 238.7, 185.2, 184.5, and 161.7 J g^{-1}, respectively. Accordingly, the latent heat increases gradually with the increase of PHC porosity. The higher porosity provides higher packaging sites for $LiNO_3$, making PHCL have more excellent latent heat, and the maximum coating ratio could reach 76.7%.

Thermal Stability of fs-CPCMs

TGA mainly determined the thermal stability of PHCL. As shown in Fig. 7, the weight loss of $LiNO_3$ and PHCL starts near 100 °C due to water, but it could still maintain a relatively stable state near its melting point. The weight-loss ratio of $LiNO_3$ was 0.5649%, and the weight-loss ratio of PHCL1, PHCL2, PHCL3, PHCL4, and PHCL5 are 0.2564, 0.4649, 0.8746, 1.8270, and 0.5483, respectively. It showed that the decomposition of $LiNO_3$ mainly caused the weight-loss rate in PHCL. This weight-loss ratio was consistent with the proportion of $LiNO_3$ contained in PHC. $LiNO_3$ was evenly distributed in ceramics, and the rest weight loss might be due to the residual oil phase and evaporation. The weight-loss rate of PHCL decreased

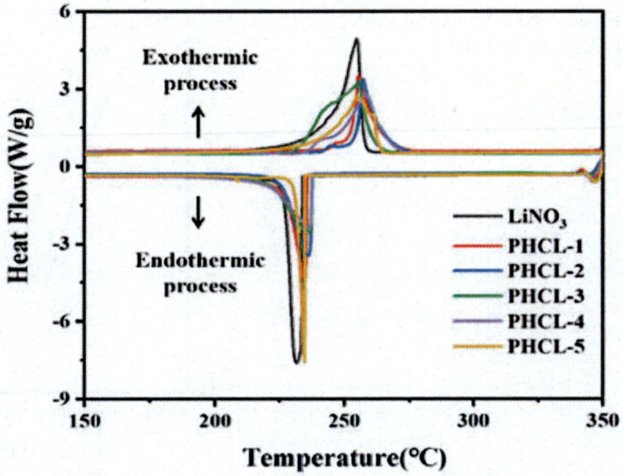

Fig. 6 DSC curves of $LiNO_3$ and the prepared PHCLs

Table 1 Thermal physical parameters of $LiNO_3$ and PHCLs

Samples	Melting process		Solidification process	
	T_M (°C)	H_M (J·g^{-1})	T_S (°C)	H_S (J g^{-1})
$LiNO_3$	254.7	307.3	231.3	317.3
PHCL1	255.7	240.2	235.8	252.2
PHCL2	256.8	221.4	234.3	238.7
PHCL3	255.1	211.2	235.1	185.2
PHCL4	257.1	192.2	236.3	184.5
PHCL5	254.5	153.5	234.2	161.7

with the increase of porosity. The sample had maintained a stable state after water decomposition and had good thermal stability, broadening its application field.

Conclusions

In this study, Hap was prepared using the Microemulsion template method. The solid content was prepared from 35 wt% to 55 wt% to form a new porous carrier with a three-dimensional aperture and window structure. The $LiNO_3$ PCM was encapsulated in the novel matrix. It had good heat storage ability and stability. The conclusion could be summarized as follows:

(1) SEM showed the unique three-dimensional ordered aperture window structure of porous ceramics prepared by the microemulsion template. At the same time, when the solid content of PHC increases from 35 wt% to 55 wt%, the apparent porosity gradually decreases from 80.18% to 55.79%. Therefore, the porous structure of PHC could be regulated by changing the solid content.

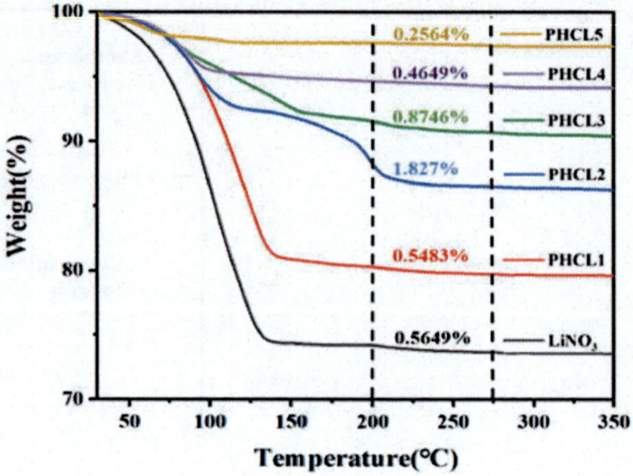

Fig. 7 TGA curves of PHC, $LiNO_3$, and PHCLs

(2) The molten impregnation of $LiNO_3$ into the pore-window structure of PHC at 1300 °C temperatures could effectively prevent the leakage of $LiNO_3$. The prepared PHC/$LiNO_3$ had a good heat storage capacity, and its enthalpy increased with the increase of porosity, up to 240.2 $J{\cdot}g^{-1}$.

Acknowledgements This work was supported by the National Natural Science Foundation of China (NSFC Grant No. U1803116). The authors also wish to thank reviewers for revising.

References

1. Da Cunha JP, Eames P (2016) Thermal energy storage for low and medium temperature applications using phase change materials—a review. Appl Energy 177:227–238
2. Nazir H, Batool M, Bolivar Osorio FJ et al (2019) Recent developments in phase change materials for energy storage applications: a review. Int J Heat Mass Transf 129:491–523
3. Deng Y, Li JH, Qian TT et al (2016) Thermal conductivity enhancement of polyethylene glycol/expanded vermiculite shape-stabilized composite phase change materials with silver nanowire for thermal energy storage. Chem Eng J 295:427–435
4. Lin Y, Jia Y, Alva G et al (2018) Review on thermal conductivity enhancement, thermal properties and applications of phase change materials in thermal energy storage. Renew Sustain Energy Rev 82:2730–2742
5. Jacob R, Liu M, Sun Y et al (2019) Characterisation of promising phase change materials for high-temperature thermal energy storage. J Energy Storage 24:100801.1–100801.12
6. Cárdenas B, Pimm AJ, Kantharaj B et al (2017) Lowering the cost of large-scale energy storage: High temperature adiabatic compressed air energy storage. Propuls Power Res 6(2):8
7. Shchukina EM, Graham M, Zheng Z et al (2018) Nanoencapsulation of phase change materials for advanced thermal energy storage systems. Chem Soc Rev 47(11):4156–4175
8. Deng Y, Li J, Qian T et al (2017) Preparation and characterization of KNO_3/diatomite shape-stabilized composite phase change material for high-temperature thermal energy storage. J Mater Sci Technol 33(2):198–203

9. Bhattacharyya S, Kumar S, Khan Y et al (2009) Biodegradable polyphosphazene nanohydroxyapatite composite nanofibers: scaffolds for bone tissue engineering. J Biomed Nanotechnol 5(1):69
10. Li Y, Li J, Feng W, Wang X, Nian H (2017) Design and Preparation of the phase change materials paraffin/porous Al_2O_3@graphitefoams with enhanced heat storage capacity and thermal conductivity. ACS Sustain Chem Eng. 5:7594–7603
11. Chang Q, Li KK, Hu SL, Dong YG, Yang JL (2016) Hydroxyapatite supported N-doped carbon quantum dots for visible-light photocatalysis. Mater Lett 175:44–47
12. Wu YF, Yang ZY, Wu NN, et al (2020) Design of three-dimensional interconnected porous hydroxyapatite ceramic-based composite phase change materials for thermal energy storage. Int J Energy Res 1–11
13. Zhao SY, Li JH, Wu YF et al (2020) Porous titanium dioxide foams: a promising carrier material for medium- and high-temperature thermal energy storage. Energy Fuels 34:8884–8890

Water Gradations Stoichiometrically Resolve Cuprous-Chloride Tetrahedral Stamps in a Hydrochloric-Acid Smelter

Kai-Wei Liu and Jia-Lin Hsu

Abstract Our proposed gradation of a bearability-changer in solution-bodied cuprous chloride interfaces the stoichiometry with the disposal-habit in precipitation. The simplicity of an antisolvent-decoupled cuprous halide in a solvent is selected for an efficacious scout around a gradational tour de force. Likewise, this select is an engineered representative to give semiconduction- and production-applicability a wieldy affordable leg up. As water specials gradationally unload the cuprous chloride near saturation in hydrochloric acid, the precipitated ones' exteriority stoichiometrically transitions. In delving an exterior-eventuality from interiority of these precipitates, their bred-in-the-bone zincblende lattice is fingered by the hired X-ray diffractometer. Optical microscopy and morphology further the visualization of the tetrahedron-externals drawn from some veterans in similar smelteries. For a constructional attraction, the escalated water bricks and downsizes an analogous façade to the primitive repeating unit at microscale. In these bricked analogies, the new discovered stellated-octahedra feature in the stoichiometric-bricking literacy and their superficial flair.

Keywords Cuprous chloride · Stellated octahedron · Antisolvent · Stoichiometry · Crystallite

K.-W. Liu (✉)
Prairie View A&M University, Mail Stop 2510, P.O. Box 519, Prairie View, TX 77446, USA
e-mail: kaliu@pvamu.edu; k-liu@tti.tamu.edu

K.-W. Liu · J.-L. Hsu
Texas A&M Transportation Institute, TAMU, 3135 TAMU, College Station, TX 77843, USA
e-mail: j-hsu@tti.tamu.edu

B. Li et al. (eds.), *Advances in Powder and Ceramic Materials Science 2023*, The Minerals, Metals & Materials Series, https://doi.org/10.1007/978-3-031-22622-9_6

Introduction

The gradability [1–5] of a reaction gray area [1–15] in a precipitation's input or output pinpoints its breadth of structural assets [1]. Plunk, the erections from our near organic-semiconductor-saturated acid are gradably commensurate with the dimensions of antisolvent imbibition [16, 17]. Hence, we posit that a degree of the antisolvent-caused scatter in the cap of liquid supportability [1, 7, 18–20] repositions components of a precipitate without another artificial kick [8, 9, 18–20]. To vet the postulated commonality, semiconductive [21–23] cuprous chloride with high hydrochloric-acid- and low water-solubility [1, 14, 24, 25] qualifies for a zincblende-lattice-halide [26–28] relay. Wholly, the black-box precipitation is manageable if an antisolvent pitch via the gradational finesse ubiquitously leans to an architectural range. Farther, assuring the postulation orchestrates a tenable paradigm for another reaction-path changer [8, 9, 18–20] incident to the quantity-resolution carriage. However, for the forthcoming advantages in this research, gradations of this solute–solvent-antisolvent team might fashion the precipitated cuprous chloride into intended architectures. To architecturally appraise a squint of crystal cuprous chloride towards out-of-bearability, our workability hinges on a compartmental water dose [1, 18, 20, 29, 30]. As reckoning water as a contributor to insolubilize a solution-borne interest, the omnipresence of water must be noticed. The hydrochloric bearer should be inevitably paced by the evaporability [31–34] thereof in the presence of atmospheric water since the sulfuric hygroscopicity [16, 17] might actuate an autonomous precipitation in the same facility. Hypothetically, proper in-solubilization of the objective crystal can be channeled. For instance, atomized droplets derail the solubilized precursor with a less rampage of intervention. On the other hand, a contact turmoil of an aqueous burden disposes the laden solute to drop far. Hither, to structurally resolve the precipitated variants, we gradate the input water for an away-from-saturation gambit. Transparentizing constructional openness unequivocally lends credence to the necessity for a gage of betweenness in a resolvent's supportability-loss variation.

Materials and Methods

General Preparation and Characterization of a Smeltery

All reagents were operated with their received conditions at the room-temperature and -humidity under the control of our facility. 0.04-g cuprous-chloride powders (97%, pure, ACROS ORGANICS) and 80-μl hydrochloric acid (($\geq$ 38%, MACRON FINE CHEMICAL) constituted a solute-acid mix. An ultrasonic bath concocted a mix in a capped glass vial for 70 s, and then, the concocted mix was transferred to a subsequent step within 5 min. A liquid in the constant volume was given by a proper pipette (Eppendorf research). A commercial humidifier was used as an atomizer and RO-water facilities supplied water. A dropped mix was slumped on its holder for 10 s in the first place. A rinse of a precipitation sequentially contained water, alcohol, and acetone as every one of them was 1 ml. The X-ray diffractometer (Cu Kα, BRUKER) and the optical (Nikon) probed interesting events to offer the crystal- and micro-structural characters. A glass substrate (VWR micro cover glass) held an interest during probing.

Video Reconnaissance

Three less-than-10-μl mixes on glass substrates contacted the combinations of the facility-environment, the atomizer, and the pipetted water doses. Their evolution was probed and recorded by the makeshift plan (Fig. 1e and [35–37]). For a fit in a format of a publication, all footage-screen-frozen snaps were just tailor-made (Fig. 1). In the snapped Fig. 1c from [37], the pavement-rate of a micro-tetrahedron was statistically reaped with a sampling area of approximately a half of 1400 μm^2.

Self-evaporation

The mix on the glass substrate was fixed by 30 μl to vaporize in the facility-environment. For an hour, the micro-structure of the smeltery was snapshotted every 5 min. For the concision of a publication, several outtakes existed in these snapshots.

Water Gradations

Any mix involved 30 μl in a glass vial. To bedrizzle a mix, the lips of the atomizer and the glass vial were connected. As manually dribbling water via the pipet, the interval among globs was less than 2 s in the scheme of gradations (Fig. 3a). Another proper pipet squirted the large-quantity water into a mix. All precipitations were centrifuged and rinsed in a sealed glass vial before the next probes. Representative images were just clipped and tailor-made (Fig. 3c).

Results and Discussion

A machination of micro-tetrahedral precipitation is prognostically glimpsed through discernible and recordable makeshifts under the optical tether (Fig. 1 and [35–37]). Janus-precipitated tetrahedra build as three cuprous-chloride-saturated hydrochloric acids moisturize under dichotomy. Harnessing an on–off switch of an atomizer and a doser stages a timing of dichotomous wetting in the environment of our facility

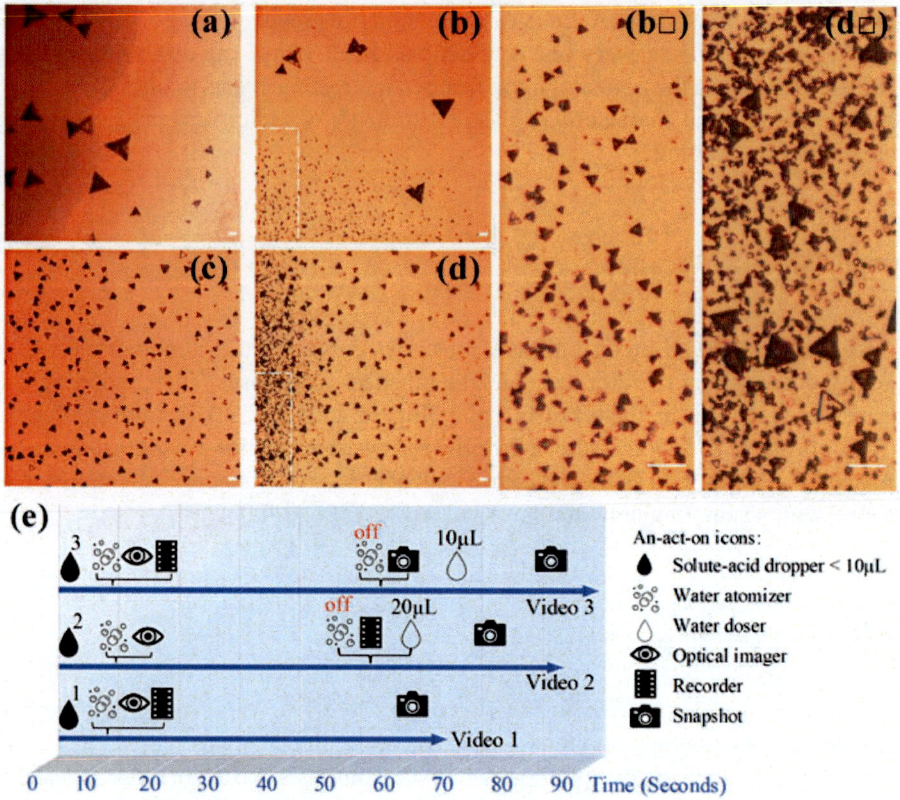

Fig. 1 The precipitated micro-tetrahedra in the juxtaposed clips of three unsized smelteries from the video reconnaissance as cuprous-chloride-hydrochloric-acid mixes on glass slides are dichotomously watered. **a** Giants via the atomizer in the smeltery 1. **b** Giants via the atomizer and then miniaturized ones via the 20-μL-water dose in the smeltery 2. **c** Giants via the atomizer in the smeltery 3. **d** Miniaturized giants via the 10-μL-water dose after **c** in smeltery 3. **e** The workflows and timing in the video reconnaissance. □ zooms in on the dashed-box selects. A rubric offs a function. The scale bar measures 50 μm

(Fig. 1e). A glass-slide-braced smelter belches a monolithic micro-tetrahedron when it is dewed. Onward, the smelter greets a volume-control-moisture injection with the pop-out midgets between their border (Fig. 1b, d from [36, 37]). On the skirts of the smelter, a merger of sweat might pertain to the hydrochloric participant and the atomized one (Fig. 1a and [35]). The offices of a jetty giant micro-tetrahedron's brilliant terminal, flip, and four-axial-stack equivalent shed its shape [35–37]. Quoad-hoc, the opaqueness of edge in a triangular monolith seems to be a terraced giant or a truncated one. For the planar span, what chalks out the stacking brake in quadrupole should be better sleuthed about. Nonetheless, our interviewees of precipitation lean to the maximization of coexistent likelihood around four poles in parqueting a tetrahedral inlay at such length-scale [35, 37]. For this maximization [35, 37], a similitude might exist in the tetrapod indium phosphide via a rapid contact of injective drops under heating [10]. A contrast between hue tinctures is how the monolithic micro-tetrahedron temporally stacks in size with a makeshift pavement-rate of a giant's culet (Fig. 1a and c from [35, 37]. Average, top, and low pavement-rate ($\mu m^2/s$) are 16.22, 25.31, and 3.17, respectively, in Fig. 1c's 50 silhouettes sans culls (e.g., drifts, rollers, and vague suspects). By comparing two water-dose starts (i.e., a mist and a 20-μl glob), the inwardness of 10-μl-discerned dwarfs of precipitation is an assurer of a graduated overhaul outward and betwixt (Fig. 1c and d from [37]). In the status quo of the video reconnaissance, albeit qualitatively, a difference in the smelted build by switching gears of a touch of water attests to our standpoint of stratification for its ambiguity of imbibing an antisolvent.

To expose a more elegant complexion of a smelter, the removal of an atomized fog achieves a simplifier of the antisolvent exertion. A binary in crystal- and micro-structure yields in a smelter's crudeness by a 30-μl vaporizable sprawl of a cuprous-chloride-hydrochloric-acid mix in the facility-air (Fig. 2). Giant-micro-tetrahedron familiars microstructurally emerge from the front of the wrestle of evaporation and facility-environment (Fig. 2a). Translucent microrods heel the familiars to loom throughout the remains during ebb and aridity of the liquid support. Cuprous chloride and cupric-chloride dihydrate inhabit crystal-structurally [14, 38–40] in the remaining familiars and microrods (Fig. 2c). However, a new micro-tetrahedron is seldom framed after approximately 30 min. This approximated watershed in the timeline can assist to surface each crystalline difference if there is a discrepancy in the solubility of the binary outcome. This discovered co-precipitation is an aid to the preciseness of similar solvent-antisolvent operations [14] as well. Namely, in operational descendants, a favorite singleton misfires if fumbling a precursor-interface resemblance.

Giant micro-tetrahedra are monumentalized by cuprous chloride in a sequential-rinse postmortem of the reactor at the 30-min finish (Fig. 2a, c). Water, alcohol, and acetone (1-ml individual) in sequence withdraw the microrods before airing the liquid residue in the monument-stood reactor. The withdrawal via water specifies the soluble microrods and an almost barren reactor. Giant-micro-tetrahedron cuprous chloride and its water-flush stationary identify the co-precipitated microrods as cupric-chloride dihydrate (Fig. 2b, c). Both the non-cuprous-chloride identity and a writ-large-descendant microrod cannot be downplayed as stepchildren. In other words, under the facility-air-humified vaporization, this smeltery is stoichiometrically and configurationally agile to lay a second. The ado about the license in precipitation often dispatches the precision and effective manipulation. However, the antisolvent agility or a precipitated complex seconds our proposed graduation of a bearer's dissipation. A virgin land of diluting a common cuprous-chloride-hydrochloric-acid mix [14] with water awaits ranking. In our stands, the birth of the non-cuprous-chloride second can be cranked out. Native cupric-chloride dihydrate in the cuprous-chloride powders under long-humidity exposure can be categorized into the minor influences. Cuprous chloride even dominates in the exposed powders because of the checks of their intact soluble behaviors in water. Unlike the native impurity, key stoichiometric ingredients inside of a flow-to-ebb bearer are responsible for the second. Downstream, a water- or a chlorine-content high of the hired hydrochloric

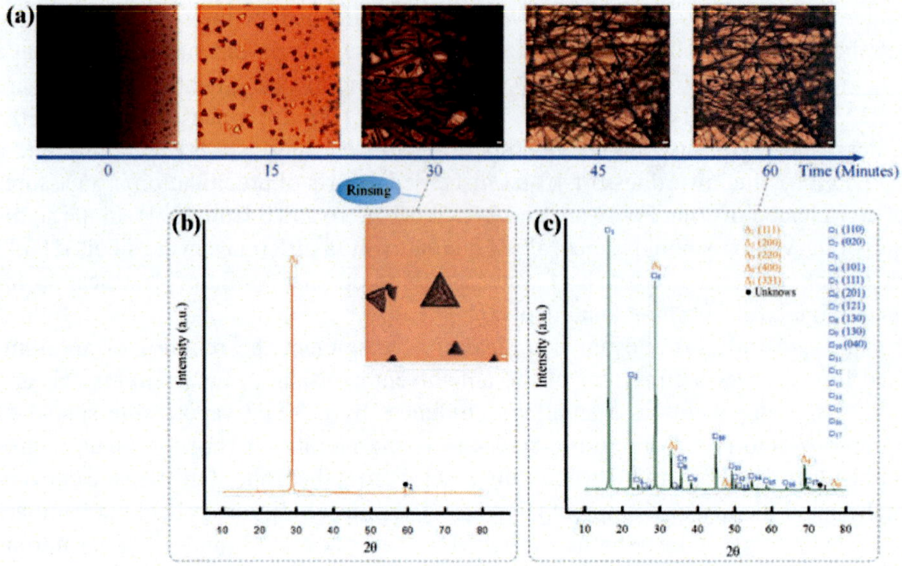

Fig. 2 An isovolumetric smelter infographic as a 30-μl vaporizable sprawl of cuprous-chloride-hydrochloric-acid mix on a glass slide evolving in the facility-air. **a** The co-precipitation of micro-tetrahedrons and -rods in the lifespan of the smelter. **b** Cuprous-chloride micro-tetrahedrons at the 30-min finish after a rinse. **c** The X-ray-diffracted coexistence of cuprous and cupric chlorides in the 60-min-old smelter. The scale bar measures 50 μm. Miller indexes tag throughout (**b**) and (**c**)

acid matches a provider of a hydrate. Because the impure cuprous-chloride powders can be refined, we would like to name this whole solute–solvent-antisolvent team as a smelter.

The quantity-resolution of the away-from-bearability watering gradationally zones vintage micro-granules in the shape-size-count map in cuprous-chloride unison (Fig. 3). For the antisolvent quantification, in a glass vial, 40 drops of 10-μl incoming water entirely titrate a 30-μl-solute-acid mix out of its olivine predecessor (Fig. 3a). A fleet repeat of this color transition stakes out the main compass in the quantitative gradation as the 0.4-ml incoming water applies (Fig. 3a). The 3-ml incoming water fringes the final status of all diluents to elude an akin trigger to metallization or oxidation from a cuprous-chloride root [14]. For a less-than-0.4-ml glob of incoming water, a one-by-one fill is dosed till the color-change adequacy and then the set dilute maximum is replenished. Oppositely, the decoloring of the mix is churned by an over-0.4-ml globule anterior to its maximum compensation. Through these titration-workflows in gradations, the initial concentration and the final one of all smelteries are the same. With this fixation of concentration, a precipitation-path is only changed by its incoming water size. If the incoming-water-size-determined paths can vary the output structures, these variations support that the dividable betweenness can be a reaction gray area in our hypothesis. After precipitating and centrifuging, a granular-and-white breed accompanies a colorless liquid unexceptionally in the

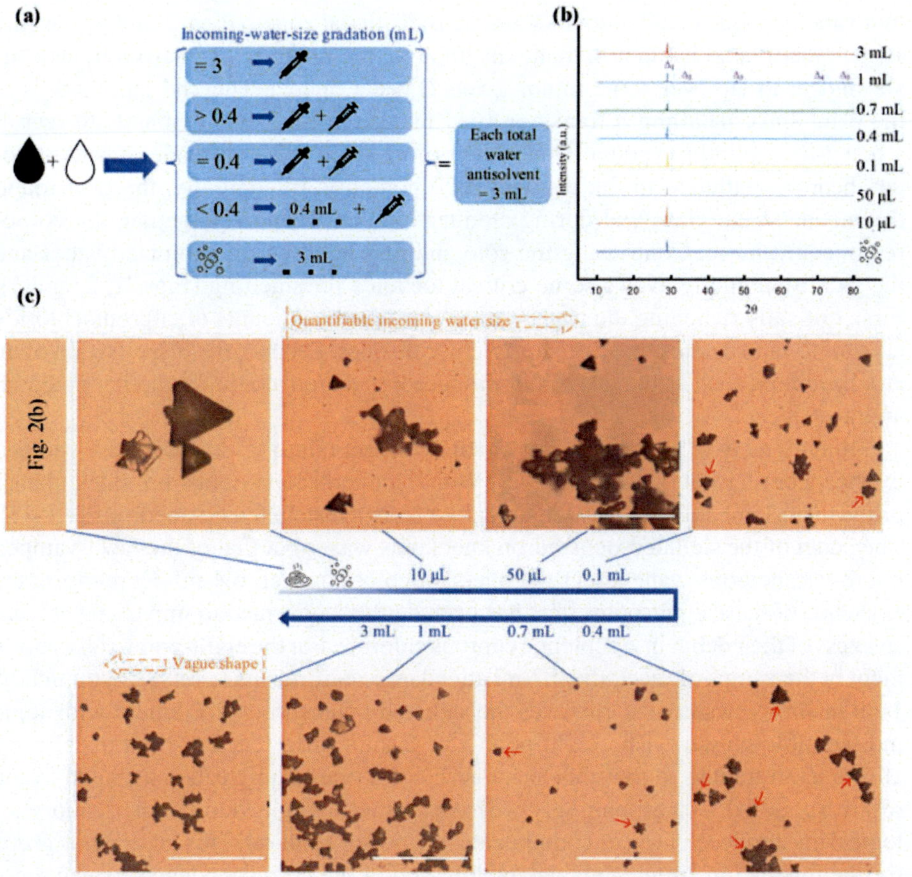

Fig. 3 A cuprous-chloride-morphology showcase in water gradations. **a** The titration-workflows in water gradations. **b** The zincblende-cuprous-chloride characteristics of the precipitations under X-ray diffraction. **c** The shape-size-count showcase of the precipitations in water-quantity-resolution axis with disproportional ticks. An-act-on icons represent 30 μL solute-acid drop, Self-evaporation, Atomization, Water dose, ··· Drop-by-drop to, The compensation to 3 mL, Single drop, and + Plus. The notations of liquid quantity denote the size of incoming or engaged water. The scale bar measures 50 μm. Rubric arrows prick examples of stellated octahedrons. Miller-index tags refer to Fig. 2

pristine produce. A zincblende streak of cuprous chloride predominantly perseverates in any X-ray-diffraction-gutted white (Fig. 3b). Profiling the resolved particulates visualizes that increasing the incoming water size tapers the particle size and the smallest-repeating-unit-analogous façade (Fig. 3c). Nevertheless, the fallibility of morphology is unavoidable in a showcased zone due to the fluent freedom of smeltery. The interfacial fluidity [34] might wander astray at a local site such that a particulate glitch is sided to the mainstream style. The preponderance of precipitation sites results in cuprous-chloride have-nots to downscale the size of a crystallite if putting a vexing glitch aside. In a dissection of 9-incoming-water-sized façades, there are 2 unequivocal landmarks among transitional zones (Figs. 2 and 3c). At the 10-μl-quantifiable landmark, manually, though, the majority of micro-tetrahedrons are midget to lap over some minor giants (Figs. 1 and 3). The 2nd landmark is at the 3 ml since detailing a transmogrified feature in this level demands to recruit a better-discernibility spotter. Under the tips of the atomized drizzle, most, micro-tetrahedrons pullulate in the equivalent corporeality. Inferentially, the dew round the mouth of the glass vial during atomization can weep to potentiate the corporeal inequivalence. Contrasting the solo and duo between the atomized water and the facility-humidity avows some critical toeholds for smelting (Figs. 1, 2, and 3). First, critically richening the water antisolvent ups the amounts of cuprous-chloride crystallites and deletes those of cupric-chloride ones. Second, the dispensation of an atomized flux, such as the gist of our proposed gradation, could be another potential architect.

From 50 μl to 0.1 ml, the facets of some midget micro-tetrahedrons are vamped by the incoming water sizes (Fig. 3c). With the microscope-optics-provided plausibility, the 0.1-ml engagement sporadically enters the stellated octahedrons [41–43]. The count of the stellated-octahedron knockouts waxes but that of the facet-vamped micro-tetrahedrons wane under the one-shot color changer, 0.4 ml. Heretofore, the existence of these eight-point stars has been double-confirmed to impart the persuasiveness of their debut in zincblende cuprous chloride. For the eight-point-star evolvement in the vamps of facet, the 0.7-ml attendance reaffirms its reality. Rhythmically, 1-ml incoming water size climaxes the stellated-octahedron vanishment with some indiscernible clusters. The eight-point-star fashion in a reaction-rhythmical zone allows us to tumble to that this shape-size-count transit might be attributed to the reactive concord. Correspondingly, we have aspired after the foundation of a morphological masterpiece via an uncomplicated reaction of solution. This meticulous grope for the foundation strengthens the exploitation of the correspondences tween these cuprous-halide-semiconductor [21–23, 26–28] stars which contours [12] capably fusillade [44, 45].

We extrapolate what antisolvent water gradations tick in the facade-engineering [1–15, 41–43] of cuprous chloride. Especially, some tetrahedron-façade-primordiality relations between the smallest unit brick and the bricked manifestations [23] can be plausibly demystified. In our extrapolation, the water-gradated smelter partakes of contact drops at the solution-antisolvent-air interface. The interfacial-mass-transport contours [6, 31–34] involve the conjunction of stoichiometric, kinetic, and contact-mechanical partakers. Through a given size of incoming water, all mass-law partakers coordinate and route their rhythm to style the shape, size, and count of precipitation. A contour characteristic of a monodisperse façade can be perfected by the goldilocks rhythm of one near ideal incoming water size, like 0.4 ml. Therefore, the observed presence of stellated octahedra peaks at the 0.4-ml-sized incoming water in the transition-zone record of their façade, roughly, from 0.1 to 0.7 ml. Reasonably, the sequence of incoming water sizes in our gradations shows several transitional zones and the stoichiometric clout thereof (Fig. 3a, c). Micro-tetrahedron, stellated octahedron, and blur stoichiometrically zone and transition their shape, size, and count. Indeed, the more sensible interdependence of them should be conducted statistically by the regions of the measurable water in the future. Based on the above, hereupon, we would like to register the reaction-mass-stoichiometry specificity between the shapes of a brick and its brickwork for a fluidically-begot crystal of a tetrahedral unit. The stoichiometric ratio between the masses of the contact fluids can be an ingenuity to ingenuously lay the smallest repeating unit into its analogous facades. De facto, all fluid-mass bricklayers rhyme to habit tetrahedron-analog facades at the very contact inception since the incoming water size determines the precipitation-path. For the solution-concretized precipitates, once stoichiometry- and contact-fluidics-closeup are considered, they aid the hired reaction-kinetics to control the process. For example, we have a hunch that tiering a supersaturated gap might be a beacon to the chirality-or-not labyrinth of tellurium [4, 5]. Additionally, regardless of the assist of ligands [4, 5, 13, 42, 43], the contact fluids might perform an unidentified rhythm of a shape-size-count transit to produce achiral and different chiral tellurium at the same go. In the sequel of the disposal-habit in the zincblende-cuprous-chloride brickfield, we keep viewing what an orthodoxy of sovereignty is. In our documented precipitation, stoichiometric and contact nuances in fluids helm the smallest motif to analogously brick up larger facades. However, holus-bolus inundated by a 3-ml torrent, delineating an individual in the major tribe of cuprous chloride, is seat-of-the-pants under the exerted optics. A workaround via cuprous-chloride cuboid in a similar smeltery to ours nudges us to allude to an antisolvent in effect for the overarching lore of a bricked facade. The 3-ml turf may allow a tinge of cubing cuprous chloride because of a large quantity of incoming water at one time but some cliffhangers are popped at us. For instance, grubbing on what dices cuprous chloride into a non-smallest-motif-analogous quirk is a must although a cubic shape is a cliché in the zincblende-tetrahedron brickfield. So far, the smelted cuprous chloride is still playing peekaboo in the entire water-quantity-resolution scale. Fortunately, at least, we can grip that unlike a fractal, the smallest-repeating-brick-analog epitome discontinues at some published points of a large quantity of incoming water at one time [14]. Invigoratingly, some intervening frontiers waiting for gradations reinforce our proposal again before reaching the first smallest-motif-analogous discontinuity's incoming water size. For the repeating-unit refusal in faceting over the discontinuity's incoming water size, we have attempted snatching an eclipse of the wetly-dwindled cuprous chloride's metallic, oxidative [14], and redissolved tendencies. These attempts favor to latch upon whether a counterpart of tetrahedra reenters the crystal shape.

Conclusions

The retrofitted façade of cuprous chloride commensurately disposed by the antisolvent gradations authenticates that a no man's land among precipitation routes necessitates to be gauged. The newfangled stellated octahedra transitioned by the coups of the gradated water render the stoichiometric concern a foothold in the keystones of the solution-routed kinetics. Zincblende-tetrahedron-analogous brickwork can be a Rosetta-stone to usher in the whys and wherefores of a hole-and-corner habit in the non-primitive-repeating-unit transmogrification.

References

1. Taylor AD et al (2021) A general approach to high-efficiency perovskite solar cells by any antisolvent. Nat Commun 12(1):1878
2. Luan C et al (2022) A real-time in situ demonstration of direct and indirect transformation pathways in CdTe magic-size clusters at room temperature. Angew Chem 134(35):e202205784
3. Vartanian, A., Some transformations happen within. Nature Reviews Materials, 2022.
4. Ben-Moshe A et al (2021) The chain of chirality transfer in tellurium nanocrystals 372(6543): 729–733
5. Popov I (2021) Is chiral crystal shape inherited or acquired? 372(6543):688–688
6. Sevim S et al (2022) Chirality transfer from a 3D macro shape to the molecular level by controlling asymmetric secondary flows. Nat Commun 13(1):1766
7. Oki O et al (2022) Synchronous assembly of chiral skeletal single-crystalline microvessels. 377(6606):673–678
8. Yang W, Yang J, Shin HS (2022) Phase- and composition-controlled synthesis. Nat Mater
9. Zhou J et al (2022) Composition and phase engineering of metal chalcogenides and phosphorous chalcogenides. Nat Mater
10. Kim Y et al (2021) Tailored growth of single-crystalline InP tetrapods. Nat Commun 12(1):4454
11. Grote L et al (2021) X-ray studies bridge the molecular and macro length scales during the emergence of CoO assemblies. Nat Commun 12(1):4429
12. Yan, C., et al., Facet-selective etching trajectories of individual semiconductor nanocrystals. 2022. Sci Adv 8(32): eabq1700.
13. Wang Y et al (2013) Shape-controlled synthesis of palladium nanocrystals: a mechanistic understanding of the evolution from octahedrons to tetrahedrons. Nano Lett 13(5):2276–2281
14. Liu H et al (2013) Scalable synthesis of hollow Cu2O nanocubes with unique optical properties via a simple hydrolysis-based approach. J Mater Chem A 1(2):302–307
15. Akkerman, Q.A., et al., Controlling the nucleation and growth kinetics of lead halide perovskite quantum dots. Science, 2022.
16. Liu K-W, Hsu J-L (2019) An innovative self-weld framework of microscale copper phthalocyanine. SCIREA Journal of Materials 4(1):1–13
17. Liu K-W, Hsu J-L (2020) α-periodicity is spontaneously phased in an acicular sulfuric-recrystallized precipitate of copper phthalocyanine. Eur J Appl Sci 8(6):81–92
18. Shi D et al (2015) Low trap-state density and long carrier diffusion in organolead trihalide perovskite single crystals 347(6221):519–522
19. Liu Y et al (2015) Two-inch-sized perovskite CH3NH3PbX3 (X = Cl, Br, I) crystals: growth and characterization 27(35):5176–5183
20. Rakita Y et al (2016) Low-temperature solution-grown CsPbBr 3 single crystals and their characterization. Cryst Growth Des 16(10):5717–5725
21. Liu A et al (2021) Engineering Copper Iodide (CuI) for multifunctional p-type transparent semiconductors and conductors 8(14):2100546
22. Gong C et al (2019) Self-confined growth of ultrathin 2D nonlayered wide-bandgap semiconductor CuBr flakes 31(36):1903580
23. Ono S (2021) Metastability relationship between two- and three-dimensional crystal structures: a case study of the Cu-based compounds. Sci Rep 11(1):14588
24. Fritz JJ (1982) Solubility of cuprous chloride in various soluble aqueous chlorides. J Chem Eng Data 27(2):188–193
25. Shang Y, Zhang D, Guo L (2011) CuCl-intermediated construction of short-range-ordered Cu2O mesoporous spheres with excellent adsorption performance. J Mater Chem 22:856–861

26. Natarajan, G., et al., Atomic layer deposition of CuCl nanoparticles. Applied Physics Letters, 2010. 97.
27. Kondo S, Kakuchi M, Saito T (2004) Photoluminescence from microcrystalline CuCl films grown from the amorphous phase. J Phys: Condens Matter 16(45):8085–8092
28. Krumpolec R et al (2018) Structural and optical properties of luminescent copper(I) chloride thin films deposited by sequentially pulsed chemical vapour deposition 8(10):369
29. Liu K et al (2022) Moisture-triggered fast crystallization enables efficient and stable perovskite solar cells. Nat Commun 13(1):4891
30. Yan Y et al (2021) Implementing an intermittent spin-coating strategy to enable bottom-up crystallization in layered halide perovskites. Nat Commun 12(1):6603
31. Guo W et al (2021) Non-associative phase separation in an evaporating droplet as a model for prebiotic compartmentalization. Nat Commun 12(1):3194
32. Andalib S, Taira K, Kavehpour HP (2021) Data-driven time-dependent state estimation for interfacial fluid mechanics in evaporating droplets. Sci Rep 11(1):13579
33. Ober P et al (2021) Liquid flow reversibly creates a macroscopic surface charge gradient. Nat Commun 12(1):4102
34. Escala DM, Muñuzuri AP (2021) A bottom-up approach to construct or deconstruct a fluid instability. Sci Rep 11(1):24368
35. Liu K-W, Hsu J-L (2022) Video 1. https://youtu.be/2JRxGW8f4jw
36. Liu K-W, Hsu J-L (2022) Video 2. https://youtube.com/shorts/ltkn7q3XGvA
37. Liu K-W, Hsu J-L (2022) Video 3. https://youtu.be/l0p5OnnRyC4
38. Han MS et al (2003) Surface properties of CuCl2/AC catalysts with various Cu contents: XRD, SEM, TG/DSC and CO-TPD analyses. Appl Surf Sci 211(1):76–81
39. Ferrandon MS et al (2010) Hydrolysis of CuCl2 in the Cu–Cl thermochemical cycle for hydrogen production: Experimental studies using a spray reactor with an ultrasonic atomizer. Int J Hydrogen Energy 35(5):1895–1904
40. Poreddy R, Engelbrekt C, Riisager A (2015) Copper oxide as efficient catalyst for oxidative dehydrogenation of alcohols with air. Catal Sci Technol 5(4):2467–2477
41. Yin Z, Lin L, Ma D (2014) Construction of Pd-based nanocatalysts for fuel cells: opportunities and challenges. Catal Sci Technol
42. King ME, Personick ML (2017) bimetallic nanoparticles with exotic facet structures via iodide-assisted reduction of palladium. 34(5):1600422
43. Villarreal E, Li GG, Wang H (2018) Carving growing nanocrystals: coupling seed-mediated growth with oxidative etching. Nanoscale 10(39):18457–18462
44. Vartanian A (2021) The many facets of cooperativity. Nat Rev Mater
45. Ye R et al (2021) Nanoscale cooperative adsorption for materials control. Nat Commun 12(1):4287

Phase Equilibria of SiO_2-Ce_2O_3-CaO-25wt.% Al_2O_3 System at 1773 K

Rensheng Li, Mengchuan Li, Tongsheng Zhang, and Wanlin Wang

Abstract The utilization of rare earth resources, especially secondary resources (e.g., RE-oxide system slag), has been limited by the lack of thermodynamic information. In order to supplement and refine the thermodynamic data related to rare earth, the equilibrium experiments of SiO_2-Ce_2O_3-CaO-25wt.% Al_2O_3 system phase diagram were carried out at 1773 K by the high-temperature isothermal equilibration/quenching technique in the current paper. The composition of seven phase regions was determined by FE-SEM, XRD, EPMA, and XRF analysis on the samples obtained by high-temperature equilibrium technology at 1773 K, including the primary crystal regions of three compounds (C_2AS, $2CaO{\cdot}SiO_2$, and $CaO{\cdot}2Ce_2O_3{\cdot}3SiO_2$), three three-phase coexistence regions (L + C_2AS + $2CaO{\cdot}SiO_2$, L + C_2AS + $CaO{\cdot}2Ce_2O_3{\cdot}3SiO_2$, and L + $CaO{\cdot}2Ce_2O_3{\cdot}3SiO_2$ + $CeAl_{11}O_{18}$), and a liquid region. The phase relations and isotherms of SiO_2-Ce_2O_3-CaO-25wt.% Al_2O_3 system obtained in current work are beneficial to the recycling of rare earth resources containing cerium.

Keywords Rare earth · Phase equilibria · Experimental thermodynamics · Liquidus

R. Li · M. Li
School of Metallurgy and Environment, Central South University, Changsha 410083, P. R. China
e-mail: rensheng_li@163.com

M. Li
e-mail: csu_lmc@163.com

T. Zhang (✉) · W. Wang
School of Metallurgy and Environment, Central South University, Changsha 410083, P. R. China
e-mail: tongsheng.zhang@csu.edu.cn

B. Li et al. (eds.), *Advances in Powder and Ceramic Materials Science 2023*, The Minerals, Metals & Materials Series, https://doi.org/10.1007/978-3-031-22622-9_7

Introduction

Generally speaking, the research methods of phase diagram include theoretical calculation and experimental determination. Theoretical calculation refers to the thermodynamic database based on strict thermodynamic data, so as to accurately describe a specific system. Therefore, to ensure the accuracy of the calculated phase diagram, a large number of reliable experimental data are necessary to determine the parameters in the thermodynamic model. So, determining the phase of some systems with practical application value through experiments is necessary.

Lanthanide elements from La to Lu as well as Y and Sc are collectively referred to as rare earth elements (REEs) due to their similar chemical properties [1]. REEs have high chemical activity and play a significant role in many new materials fields. In particular, the improvement of mechanical properties of ceramic materials and steel materials by REEs has attracted more and more attention [2]. In recent years, with the global consumption of high-grade rare earth resources increasing significantly, the recycling of waste containing rare earth has been brought into focus [3].

As we all know, the Bayan Obo deposit in Inner Mongolia, North China is rich in rare earth, titanium, iron, fluoride, and other precious resources [4]. However, the associated beneficiation and smelting processes are very hard because it is a typical polymetallic ore, and valuable components coexist [5, 6]. Besides, due to the limitations of current technologies, the utilization rate of rare earth resources is very low. Only about 10% of the rare earth resources in the world are utilized, which is out of proportion to their value [7]. More importantly, with the accumulation of tailings dams, many serious environmental problems such as water pollution have emerged. Therefore, the effective recovery and application of rare earth elements are of great significance to the sustainable utilization of rare earth slag [8].

Recently, many scholars have studied and reported a great number of phase equilibrium data of binary and ternary systems in rare earth slag systems. Tas et al. [9] studied Al_2O_3-Ce_2O_3 system, and found that there are two binary compounds ($CeAl_{11}O_{18}$ and $CeAlO_3$) in the system, and provided lattice constants of $CeAl_{11}O_{18}$. At 1873 K, two solid solution phases ($Ce_{9.33-x}Ca_x(SiO_2)_6O_{2-\delta}$ and $Ca_{2-x}Ce_xSiO_{4+\delta}$) were found in SiO_2-Ce_2O_3-CaO system, and the formulae for calculating the activity of Ce_2O_3 are provided [10]. Lan et al. [11, 12] provided the phase diagrams of SiO_2-Ce_2O_3-CaO system at 1573 and 1773 K, and the phase diagrams of CaO-SiO_2-CaF_2-Ce_2O_3 system at 1373 K. As we all know, P and F are also the main compositions in rare earth ores, and they are transferred to the slag containing rare earth, which affects the composition and performance of the slag. Therefore, Lan constructed an isothermal pseudo-ternary phase diagram of CaO-SiO_2-CaF_2 (30wt. %)-P_2O_5 (10wt. %)-Ce_2O_3 and obtained the formation mechanism of boro-magnesium olivine in the five-member system [13].

Recently, many different phases of REEs have been found in slag containing rare earth elements in previous studies. By analyzing and simulating the morphology of slag containing rare earth elements, Li et al. [14] found that rare earth elements are converted into calcium-cerite phase. Ding et al. [15] reported that rare earth elements mainly exist in the form of cephalosilicate in the rare earth slag produced from Bayan Obo ore. Therefore, the supplement and improvement of thermodynamic information can also help researchers to confirm that the rare earth phase is a series of component transformation behaviors according to the formation and transformation information of rare earth phase in equilibrium state. On this basis, the equilibrium phase composition of SiO_2-Ce_2O_3-CaO-25wt.% Al_2O_3 at 1773 K was studied in this paper, which provides basic phase transition data for the recycling of rare earth waste residue in ceramics and other fields.

Methodology

Rare Earth Slag Samples Preparation

Generally speaking, isothermal pseudo-ternary phase diagrams can be constructed by fixing typical components, so as to simplify the study of phase equilibrium in multicomponent systems. In order to obtain uniform composition and ensure the accuracy of the experiment, firstly, the SiO_2, CaO, CeO_2, and Al_2O_3 powders of 99.99% purity (Sinopharm Chemical Reagents Co., Ltd.) were put into a muffle furnace and calcined at 1273 K for 6 h to remove moisture and impurities. The raw materials were accurately weighed by an electronic balance with an accuracy of 0.0001 g, and then used to make slags. Every slag was then ground by an agate mortar for 30 min, so that it can be fully and uniformly mixed. Thereafter, the pre-mixed slags were put into a vertical furnace (temperature measurement accuracy $\pm$ 1 K) with Ar + 10% H_2 atmosphere. The powdered slags in platinum crucibles were heated to 1923 K for 2 h. Then, the molten samples were quickly quenched in the ice-water mixture. In current study, it is a key point to ensure the complete transformation of CeO_2 into Ce_2O_3. Therefore, we referred to a large number of literatures [9–13] and adopted the same reduce atmosphere. The oxygen partial pressure was controlled below 2 $\times 10^{-16}$ during the experiment process. Based on these experimental conditions, we also conducted X-ray Photoelectron Spectroscopy (XPS) test on all of the samples. Taking sample N1 for example, the $3d_{5/2}$ and $3d_{3/2}$ XPS lines of the Ce^{3+} ion is presented in Fig. 1. Because of the transition behavior between different energy levels of Ce^{3+} ion, the $3d_{5/2}$ doublet and the $3d_{3/2}$ doublet appear [16, 17]. The results showed that only Ce^{3+} but no Ce^{4+} existed in the pre-melting samples.

X-ray diffraction (XRD, D/max 2550VB) detection is required for all slag samples to ensure the uniform glass phase after quenching. If crystal precipitation was detected in the sample, the sample would be reproduced for the supplementary experiment. The content of each oxide will be readjusted, and the slag samples pre-melting test will be conducted again until all the required quenched samples completely show the glass phase. Figure 2 shows the microstructure (a) and XRD patterns (b) of the glass phase sample N1.

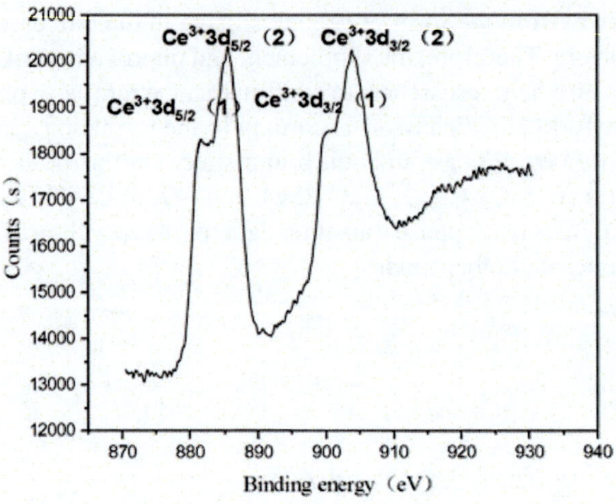

Fig. 1 XPS detection map of quenching slag sample N1

Because of burning loss and batching error, the composition of quenching slags may be different from that of samples, so it was necessary to detect the composition of quenching slags to determine the composition of slag samples again to ensure the accuracy of the experiment. X-Ray Fluorescence Spectrometer (XRF, Axios mAX) was used to analyze the composition of quenching slag samples, and the results are listed in the pseudo ternary phase diagram of SiO_2-Ce_2O_3-CaO-25wt. % Al_2O_3 system, as shown in Fig. 3.

Equilibration Experiments

In the equilibration experiments, a box furnace, as shown in Fig. 4a, was used, and the temperature was detected by a B-type thermocouple placed at the bottom of the furnace. The overall temperature accuracy is $\pm$ 1 K. High-purity argon with 10% hydrogen, purified by gas purification unit as shown in Fig. 4b, was continuously inject as protective gas to avoid the oxidation of slag samples and isolated other possible pollution sources. The deoxidation unit consists of a heating device, and a deoxidation chamber with copper, sponge titanium, and magnesium, which can keep the partial pressure of oxygen in argon gas below 10^{-16}. About 2 g of homogenized glass-phase slag sample was loaded into a platinum crucible and placed in the constant temperature zone of the box furnace. All samples were heated to 1923 K again and held for 30 min, then cooled to the target equilibrium temperature (1773 K) at a rapid rate (10 K/min). Based on preliminary experimental results and previous experimental experience [18], the time was set at 24 h to ensure that the equilibrium is achieved.

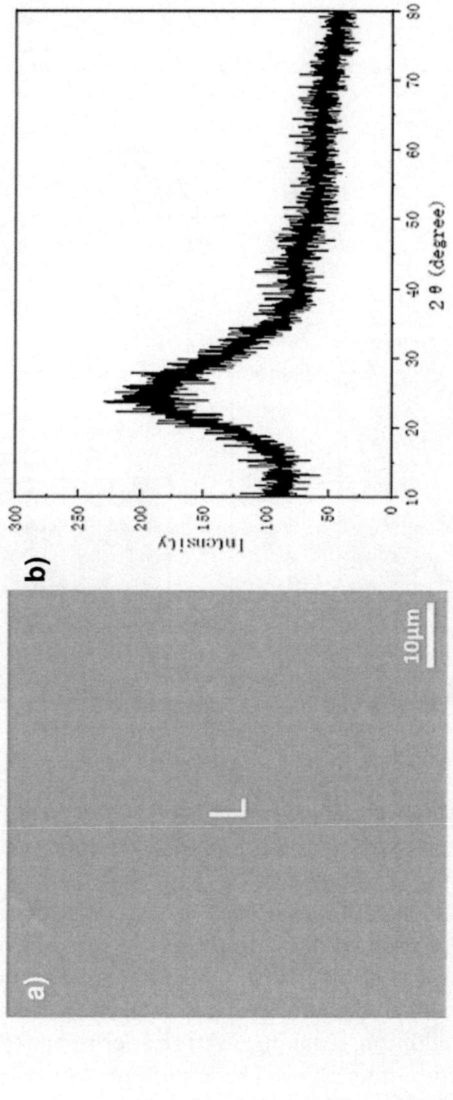

Fig. 2 Microstructure (**a**) and XRD patterns (**b**) of pre-melted slag sample N1

Fig. 3 Pre-melting component points of SiO_2-Ce_2O_3-CaO-25wt. % Al_2O_3 system at 1773 K

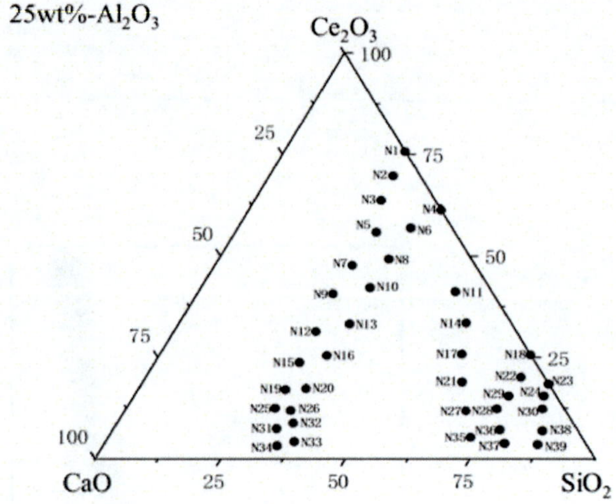

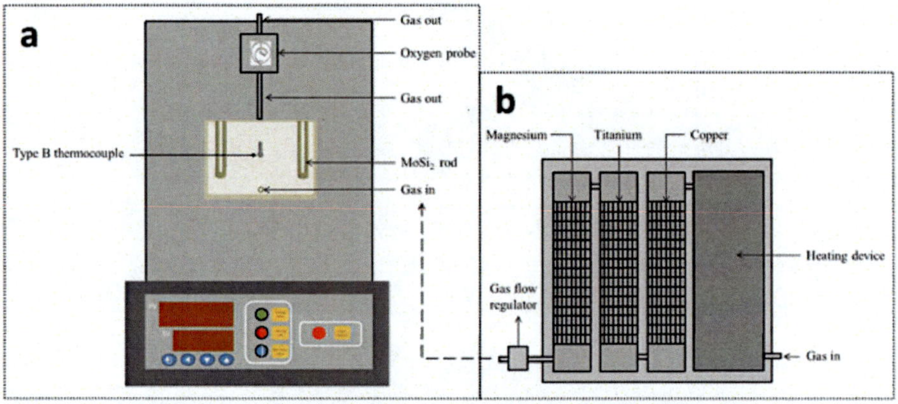

Fig. 4 Schematic diagram of box-type furnace (**a**) and gas purification unit (**b**)

When the samples reached equilibrium, they were rapidly quenched into the ice-water mixture. Then the quenched slag samples were dried at room temperature. Electron probe microanalysis (EPMA, JXA-8230) was used to analyze the composition of each sample, and the precipitated phase contained in hardened slag samples was detected by XRD. In addition, scanning electron microscope (SEM, JSM-7900F) was used to observe the coexisting phase. The equilibrium micro-morphology of the samples was obtained in BSE mode. SEM test conditions are: beam current 15 mA, electron beam acceleration voltage 20 kV. The EPMA measurement conditions were as follows: a beam current of 10^{-8} A, an accelerating potential of 20 kV, and a probe diameter of 1 μm. $CaSiO_3$, Al_2O_3, and CeP_5O_{14} were respectively used as standards to analyze CaO and SiO_2, Al_2O_3, Ce_2O_3. For each balancing stage, more than 10 different points were analyzed and average values were calculated. Combining the results of XRD, FE-SEM, and EPMA, the phase composition of those samples in equilibrium state was obtained.

Results and Discussion

Equilibrium Phase Composition of SiO_2-Ce_2O_3-CaO-25wt.% Al_2O_3System at 1773 K

In general, by fixing typical components to construct isothermal pseudo-ternary phase diagram, the investigation related to the multicomponent system phase equilibrium can be simplified [19]. When quenched to room temperature, some samples would have a large amount of secondary precipitation, which leads to inaccurate liquidus composition and is an important reason for the failure of the experiment. Therefore, some samples were selected to study the phase equilibrium of SiO_2-Ce_2O_3-CaO-25wt.% Al_2O_3 system. The results of FE-SEM and XRD show that the quenched samples consist of amorphous glass and crystalline phase. The crystalline phase is precipitated from the solid phase at 1773 K, while the glassy phase is formed by quenching and supercooling the liquid phase in equilibrium with the solid phase.

The results of EPMA and XRD, which measured the chemical composition of glass phase and crystal phase, show that there are seven different equilibrium phases. Those seven equilibrium phases in order are: (1) the simple liquid phase; (2) the three-phase equilibrium between C_2AS ($2CaO{\cdot}Al_2O_3{\cdot}SiO_2$), $2CaO{\cdot}SiO_2$, and liquid phase; (3) the solid–liquid coexistence of C_2AS and liquid phase; (4) the three-phase equilibrium between C_2AS, $CaO{\cdot}2Ce_2O_3{\cdot}3SiO_2$, and liquid phase; (5) the solid–liquid coexistence of $CaO{\cdot}2Ce_2O_3{\cdot}3SiO_2$ and liquid phase; (6) the three-phase equilibrium between $CaO{\cdot}2Ce_2O_3{\cdot}3SiO_2$, $CeAl_{11}O_{18}$ and liquid phase; (7) the solid–liquid coexistence of $3Al_2O_3{\cdot}2SiO_2$ and liquid phase. Due to the replacement mechanism of Ca and REEs, Ce^{3+} will replace a part of Ca^{2+}, which cause that about 3 – 5wt% of CaO are replaced by REEs in the sorosilicate phase [20]. Therefore, it is remarkable that Ce^{3+} can replace part of Ca^{2+} in the compounds Ca_2SiO_4 and C_2AS, resulting in $(Ca,Ce)_2SiO_4$ and $2(Ca,Ce)O{\cdot}Al_2O_3{\cdot}SiO_2$. Gao [21] found the solid solution $CaO{\cdot}2Ce_2O_3{\cdot}3SiO_2$ during the reduction process, whose research results are the same as these in current study. The microstructure and composition of the equilibrium phase at 1773 K are shown in Fig. 5 and Table 1, respectively. As is known to all, the liquid phase area gradually expands as the temperature increases.

The phase structure of sample N33 at 1773 K is shown in Fig. 5a, showing three-phase coexistence. Combined with the detection results of EPMA and XRD, it can be found that the dark gray region is calcium aluminite melilite phase

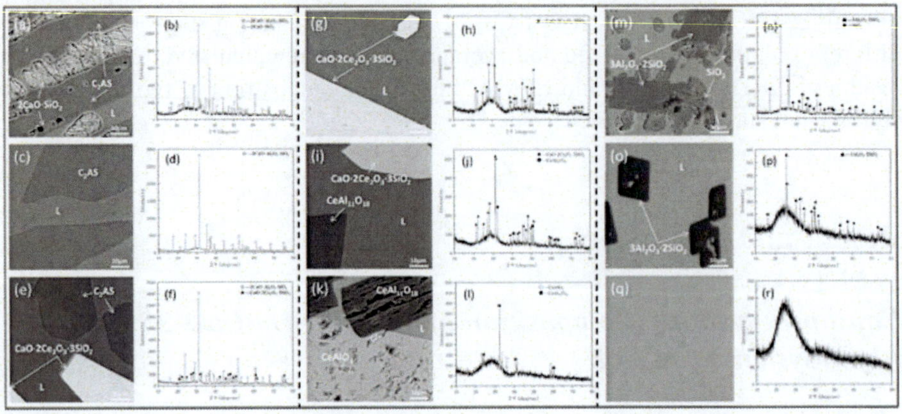

Fig. 5 Microstructure of equilibrium phase of SiO_2-Ce_2O_3-CaO-25wt. % Al_2O_3 system at 1773 K and corresponding XRD detection patterns. **a**, **c**, **e**, **g**, **i**, **k**, **m**, **o**, and **q** represent the Micromorphology of N33, N26, N9, N3, N2, N1, N18, N24, and N11 sample, respectively; and **b**, **d**, **f**, **h**, **j**, **l**, **n**, **p**, and **r** were XRD patterns of corresponding sample

($2CaO{\cdot}Al_2O_3{\cdot}SiO_2$, C_2AS), the gray region is quenched liquid phase, and the light gray region is Mosaic ($2CaO{\cdot}SiO_2$). For samples N32 and N34, the same precipitated phase, calcium aluminite melilite phase, appears, as shown in Fig. 5c. According to EPMA and XRD results, the gray region is the quenching liquid phase.

The microstructure of sample N9 is shown in Fig. 5e. According to the BSE picture, it can be clearly found that the equilibrium phase presents a state of three-phase coexistence. According to EPMA and XRD results, it can be easily confirmed that the dark gray region is still C_2AS, and the gray region is quenched liquid phase. The most recent white region can be identified as $CaO{\cdot}2Ce_2O_3{\cdot}3SiO_2$ phase. In addition, $CaO{\cdot}2Ce_2O_3{\cdot}3SiO_2$ phase also exists in samples N3, N5, and N10. Their microstructures are shown in Fig. 5g and 5(i). At 1773 K, there are only two equilibrium phases $CaO{\cdot}2Ce_2O_3{\cdot}3SiO_2$ phase and quenched liquid phase in samples N5 and N10. The results indicate that the composition of slag sample N5 and slag sample N10 is in the primary crystal region of $CaO{\cdot}2Ce_2O_3{\cdot}3SiO_2$, and there are grey $CaO{\cdot}2Ce_2O_3{\cdot}3SiO_2$ phase and light gray quenching liquid phase in sample N2, as well as dark gray long $CeAl_{11}O_{18}$ phase. For sample N1, the microstructure is shown in Fig. 5k, and the $CeAl_{11}O_{18}$ phase shows a black band. The white area can be identified as $CeAlO_3$ phase according to the XRD and EPMA detection results.

Table 1 Equilibrium phase compositions at 1773 K

Sample No	Phase	Composition (wt%)			
		Al_2O_3	SiO_2	Ce_2O_3	CaO
N1	L	25.25	26.13	48.62	0
	$CeAl_{11}O_{18}$	82.1	0.62	16.85	0.43
	$CeAlO_3$	32.72	0.43	66.43	0.42
N2	L	27.46	29.25	39.21	4.08
	$CeAl_{11}O_{18}$	82.18	0.89	16.28	0.65
	$CaO{\cdot}2Ce_2O_3{\cdot}3SiO_2$	1.59	22.27	69.57	6.57
N3	L	28.26	27.90	32.68	11.16
	$CeAl_{11}O_{18}$	82.26	0.48	16.76	0.5
	$CaO{\cdot}2Ce_2O_3{\cdot}3SiO_2$	1.29	22.08	70.16	6.47
N5	L	30.49	27.41	26.17	15.92
	$CaO{\cdot}2Ce_2O_3{\cdot}3SiO_2$	1.14	22.12	70.19	6.55
N7*	L	32.18	25.11	21.32	21.39
	C_2AS	37.13	21.1	5.93	35.84
	$CaO{\cdot}2Ce_2O_3{\cdot}3SiO_2$	1.32	22.13	70.16	6.39
N9*	L	31.83	25.88	20.98	21.31
	C_2AS	37.09	20.58	5.95	36.38
	$CaO{\cdot}2Ce_2O_3{\cdot}3SiO_2$	1.65	21.8	68.94	7.61
N10	L	32.37	26.69	22.09	18.85
	$CaO{\cdot}2Ce_2O_3{\cdot}3SiO_2$	1.08	22.59	68.93	7.4
N12*	L	22.44	32.18	16.12	29.26
	C_2AS	36.9	22.11	5.13	35.86
	$CaO{\cdot}2Ce_2O_3{\cdot}3SiO_2$	1.2	21.42	70.31	7.07
N13*	L	28.93	29.32	18.14	23.61
	C_2AS	38.28	20.12	4.43	37.17
	$CaO{\cdot}2Ce_2O_3{\cdot}3SiO_2$	0.98	21.08	70.99	6.95
N16*	L	24.37	32.45	15.90	27.28
	C_2AS	38.63	20.89	3.87	36.61
	$CaO{\cdot}2Ce_2O_3{\cdot}3SiO_2$	1.41	20.82	70.75	7.02
N20*	L	19.79	33.77	14.85	31.59
	C_2AS	37.55	22.24	3.15	37.06
	$CaO{\cdot}2Ce_2O_3{\cdot}3SiO_2$	0.87	21.67	70.52	6.94
N23	L	20.78	52.61	26.61	0
	$3Al_2O_3{\cdot}2SiO_2$	70.83	25.83	2.42	0.92

(continued)

Table 1 (continued)

Sample No	Phase	Composition (wt%)			
		Al_2O_3	SiO_2	Ce_2O_3	CaO
	SiO_2	0.02	99.68	0.14	0.16
N24	L	18.03	64.56	11.85	5.56
	$3Al_2O_3{\cdot}2SiO_2$	71.43	25.13	2.61	0.83
N26*	L	13.50	34.57	14.43	37.50
	C_2AS	37.86	20.57	3.93	37.64
	$CaO{\cdot}2Ce_2O_3{\cdot}3SiO_2$	0.22	22.9	69.71	7.17
N30	L	21.87	57.71	16.89	3.53
	$3Al_2O_3{\cdot}2SiO_2$	71.41	25.34	2.4	0.85
N32*	L	14.29	35.84	10.88	38.99
	C_2AS	38.32	21.65	4.42	35.61
N33*	L	12.72	35.51	3.75	48.02
	C_2AS	38.94	20.63	3.8	36.63
	$2CaO{\cdot}SiO_2$	1.54	32.23	6.14	60.09
N34*	L	15.38	36.58	6.78	41.26
	C_2AS	38.39	19.76	6.69	35.16
N38	L	21.05	64.84	6.80	7.31
	$3Al_2O_3{\cdot}2SiO_2$	71.65	25.22	2.6	0.53
N39	L	21.99	64.65	3.74	9.62
	$3Al_2O_3{\cdot}2SiO_2$	72.17	24.93	2.38	0.52

*represents the results obtained by converting Ce^{3+} into Ca^{2+}

The microstructure of the equilibrium phase of sample N18 is shown in Fig. 5m. According to the results of EPMA and XRD, it can be known that the round, oval, and conical phase is SiO_2 phase, while the regular quadrilateral phase is $3Al_2O_3{\cdot}2SiO_2$ phase. Similarly, in the samples N24 and N30, the relatively regular quadrilateral $3Al_2O_3{\cdot}2SiO_2$ phase is detected, but their equilibrium phase structure only contained $3Al_2O_3{\cdot}2SiO_2$ phase and quenching liquid phase. Samples N11, N14, N17, N21, and N27 only contain quenched liquid phase, and it can be determined that there is no other precipitated phase according to the XRD detection results. Figure 5q and 5(r) are the microscopic morphology images and XRD detection patterns of quenched slag sample N14. It is worth noting that in the SiO_2-Ce_2O_3-CaO-5wt.% Al_2O_3 system previously studied [18], the initial crystal region of SiO_2 appears when the content of SiO_2 is higher than 70wt%. However, in this system, when the content of Al_2O_3 is increased from a fixed 5wt% to 25wt%, no initial crystal region of SiO_2 is found.

Table 1 lists the composition and content of liquid phase and precipitates of each sample. It should be noted that part of Ca^{2+} was replaced by Ce^{3+} in the sample with ' *'. Thus, the precipitated phase Ca_2SiO_4 and C_2AS were obtained by converting Ce^{3+} into Ca^{2+}.

Determination of the Isotherms

The composition of the quenched liquid phase was standardized so that the phase relationship of SiO_2-Ce_2O_3-CaO-25wt.% Al_2O_3 system can be represented by phase diagram [22, 23]. As shown in Fig. 6, the liquid phase isotherms of SiO_2-Ce_2O_3-CaO-25wt.% Al_2O_3 system at 1773 K are obtained based on the above experimental results.

As can be seen from the Fig. 6, the liquid phase region is concentrated in the SiO_2 content between about 38% and 75%. Meanwhile, with the increase of CaO content, the distance between the liquid phase isotherms of 1773 K is closer and closer, indicating that when the ratio of SiO_2 and Ce_2O_3 content is constant, the melting point of the mixture can be significantly increased by increasing the content of CaO in the mixture. And when the binary basicity (CaO/SiO_2) is certain, reducing the content of Ce_2O_3 can also achieve the effect of reducing the melting point of the mixture. Meanwhile, with the decrease of Ce_2O_3 content, the liquid phase region becomes larger. In the previous work [18], we studied the SiO_2-Ce_2O_3-CaO-5wt.% Al_2O_3 system and not found a liquid phase region at 1773 K at the phase boundary of SiO_2 and Ce_2O_3. While in this study, there was a large liquid phase region at the phase boundary of SiO_2 and Ce_2O_3 [18]. Therefore, we can speculate that in the ternary system of SiO_2-Ce_2O_3-Al_2O_3, there is a large liquid region at 1773 K with the increase of Al_2O_3.

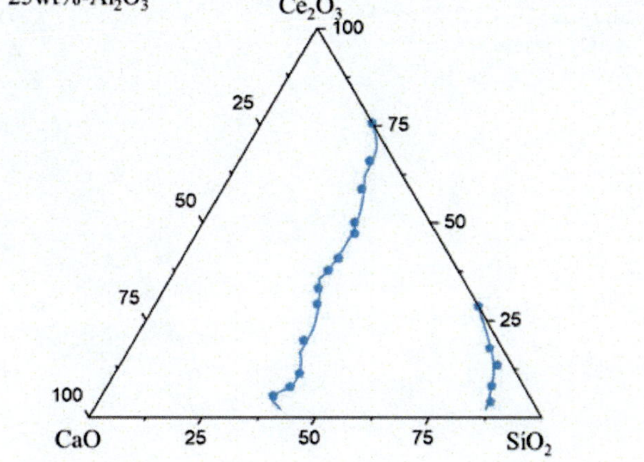

Fig. 6 Liquid phase isotherms of SiO_2-Ce_2O_3-CaO-25wt.% Al_2O_3 system at 1773 K

Construction of SiO_2-Ce_2O_3-CaO-25wt.% Al_2O_3 System Phase Diagram

Phase diagram can be used to quantitatively describe the phase equilibrium of systems. It can help to determine the direction of a reaction and avoid the resource consumption during the research process. Combining the isotherms of SiO_2-Ce_2O_3-CaO-25wt.% Al_2O_3 system at 1773 K with the above equilibrium phase structure, a phase diagram related to the system containing rare earth was constructed, as shown in Fig. 7. In Fig. 7, according to the trend of phase boundary, the intersection point of phase boundary is found. Meanwhile, several points on the CaO-SiO_2 phase boundary are data points obtained according to the known phase diagram [24].

In this study, there are six regions where equilibrium phase and liquid phase coexist, among which there are three single-phase regions, the primary phases are mullite phase, $CaO{\cdot}2Ce_2O_3{\cdot}3SiO_2$ phase, and calcium anorthite solid solution phase, and the other three are three-phase coexistence regions. From the equilibrium phase detected in 1773 K equilibrium slag samples, it can be seen that when SiO_2 content is higher than 75%, there is only $3Al_2O_3{\cdot}2SiO_2$ single phase zone. It is found that $CaO{\cdot}2Ce_2O_3{\cdot}3SiO_2$ is a favorable stable phase of RE elements according to the phase of the system. The synthesis range of $CaO{\cdot}2Ce_2O_3{\cdot}3SiO_2$ is quite wide and is greatly affected by temperature and slag composition, which indicates that a variety of components can be formed in the quaternary system. Therefore, according to the phase diagram, rare earth elements can be enriched into the $CaO{\cdot}2Ce_2O_3{\cdot}3SiO_2$ phase by controlling the appropriate composition of the slag containing rare earth elements. The current research provides indispensable thermodynamic data for RE-containing phase equilibrium, which is very important for sustainable recovery of REEs from REEs-containing slag. Additionally, the shaded area has not involved in current research, as shown in Fig. 7.

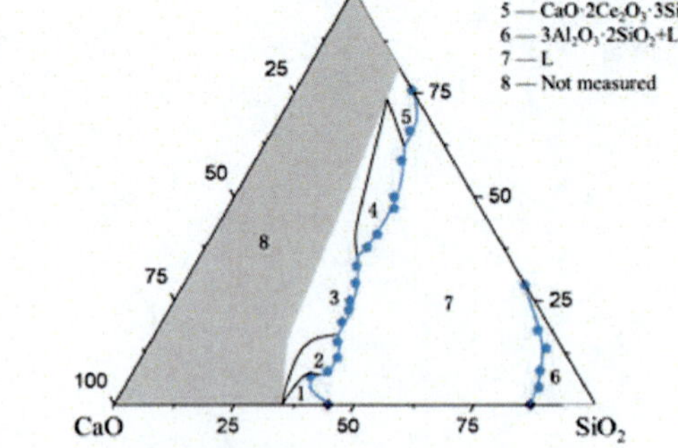

Fig. 7 Phase diagram of SiO_2-Ce_2O_3-CaO-25wt.% Al_2O_3 system

Conclusions

The thermodynamic information plays a significant role in the recovery and utilization of slag systems containing rare earth elements and also establishes a crucial theoretical basis for the development of REEs application industry. And it is also beneficial to the development and application of rare earth in iron and steel industry and ceramic industry. The main results of the phase equilibrium relationship of SiO_2-Ce_2O_3-CaO-25wt.% Al_2O_3 system phase diagram at 1773 K investigated in current research can be summarized as follows:

(1) Seven phases were determined by the current research work. It consists of a single liquid region, three primary crystal regions (C_2AS, $2CaO{\cdot}SiO_2$, $CaO{\cdot}2Ce_2O_3{\cdot}3SiO_2$) and three three-phase coexistence regions (L + C_2AS + $2CaO{\cdot}SiO_2$, L + C_2AS + $CaO{\cdot}2Ce_2O_3{\cdot}3SiO_2$, L + $CaO{\cdot}2Ce_2O_3{\cdot}3SiO_2$ + $CeAl_{11}O_{18}$).
(2) The liquid isotherms and the phase relationship of SiO_2-Ce_2O_3-CaO-25wt.% Al_2O_3 system related to the cerium-containing slag system are established. According to the phase diagram, rare earth elements could be enriched into the $CaO{\cdot}2Ce_2O_3{\cdot}3SiO_2$ phase as much as possible by controlling the appropriate composition of the slag containing rare earth elements. The phase equilibrium data can provide a basis for comprehensive utilization of cerium-rare earth slag and optimization of related thermodynamic database.

References

1. Mohammed AA, Leonard Z, Michael GP (2012) Economics of rare earth elements in ceramic capacitors. Ceram Int 38:6091–6098
2. Xu CH, Huang CZ, Ai X (2006) Toughening and strengthening of advanced ceramics with rare earth additives. Ceram Int 32:423–429
3. Sadeq MS, Morshidy HY (2019) Effect of mixed rare-earth ions on the structural and optical properties of some borate glasses. Ceram Int 45:18327–18332
4. Yuan ZX, Bai G, Wu CY, Zhang ZQ, Ye XJ (1992) Geological features and genesis of the Bayan Obo REE ore deposit. Inner Mongolia China Appl Geochem 7:429–442
5. Zhang B, Liu CJ, Li CL, Jiang MF (2014) A novel approach for recovery of rare earths and niobium from Bayan Obo tailings. Miner Eng 65:17–23
6. Jordens A, Cheng YP, Waters KE (2013) A review of the beneficiation of rare earth element bearing minerals. Miner Eng 41:97–114
7. Humphries M (2010) Rare earth element: the global supply chain. Congressional Research Service, Washington, DC
8. Xu T, Peng HQ (2009) Formation cause, composition analysis and comprehensive utilization of rare earth solid wastes. J Rare earth 27:1096
9. Tas AC, Akinc M (1994) Phase relations in the system Ce_2O_3-$Ce_2Si_2O_7$ in the temperature range 1150°C to 1970°C in reducing and inert atmospheres. J Am Ceram Soc 77:2953–2960
10. Zhao Z, Chen X, Glaser B, Yan B (2019) Experimental study on the thermodynamics of the CaO-SiO_2-Ce_2O_3 system at 1873 K. Metall Mater Trans B 50:395–406
11. Lan X, Gao JT, Du Y, Guo ZC (2020) Thermodynamics and crystallization kinetics of REEs in CaO-SiO_2-Ce_2O_3 system. J Am Ceram Soc 103:2845–2858

12. Lan X, Gao JT, Li Y, Guo ZC (2021) Thermodynamics and kinetics of REEs in $CaO-SiO_2-CaF_2-Ce_2O_3$ system: a theoretical basis toward sustainable utilization of REEs in REE-bearing slag. Ceram Int 47:6130–6138
13. Lan X, Gao JT, Li Y, Guo ZC (2021) Phase equilibria of $CaO–SiO_2–CaF_2–P_2O_5–Ce_2O_3$ system and formation mechanism of britholite. Ceram Int 47:11966–11972
14. Li DG, Bu Q, Lou T, Sui ZT (2004) Morphology of solidified slag for RE_2O_3-CaO-SiO_2-CaF_2-MgO-Al_2O_3 system. J Iron Steel Res Int 16:30–33
15. Ding YG, Wang JS, Wang G, Xue QG (2012) Innovative methodology for separating of rare earth and iron from Bayan Obo complex iron ore. ISIJ Int 52:1772–1777
16. Wu Y, Ren G, Ding D, Yang F, Pan S (2012) Energy levels of Ce^{3+} in $Lu_{0.8}Sc_{0.2}BO_3$ host: a comparison study between x-ray photoelectron spectroscopy and pure optical method. IEEE Trans Nucl Sci 59:2069–2073
17. Pidol L, Viana B, Kahn-Harari A, Galtayries A, Bessiere A, Dorenbos P (2004) Optical properties and energy levels of Ce^{3+} in lutetium pyrosilicate scintillator crystal. J Appl Phys 95:7731–7737
18. Li M, Zhang T, Wang W, Zhang H, Li R. Experimental study on thermodynamics of CaO-SiO2-Ce_2O_3 -5wt.% Al_2O_3 system at 1773 K. TMS (2022)209–223
19. Won K, Smith A, Zeininger G (2006) Thermodynamic methods for pressure relief system design parameters. Fluid Phase Equilib 241:41–50
20. Elwert T, Goldmann D, Schirmer T, Strauss K (2014) Affinity of rare earth elements to silico-phosphate phases in the system Al_2O_3-CaO-MgO-P_2O_5-SiO_2. Chem Ing Tech 86:840–847
21. Gao P, Han Y, Sun Y, Chen C (2012) Research on the reaction behavior of rare earth elements in reduction. Powder Technol Appl IV 454:268–272
22. Wang Z, Sun H, Zhang L, Zhu Q (2016) Phase equilibria in the TiO2 rich part of the TiO_2-CaO-SiO_2-10wt% Al_2O_3 system at 1773 K and 1873 K. J Alloy Compd 671:137–143
23. Gran J, Wang Y, Sichen D (2011) Experimental determination of the liquidus in the high basicity region in the Al_2O_3(30 mass%)-CaO-MgO-SiO_2 system. Calphad 35:249–254
24. Verein Deutscher Eisenhuttenleute (1995) SLAG ATLAS 2nd Edition[M]. Verlag Stahleisen GmbH, p 105

Printed Carbon Nanotube and Graphene Heaters for Drying Ceramics

Ziyad Sherif, John Patsavellas, and Konstantinos Salonitis

Abstract The ceramic manufacturing process has been subject to many advances with the evolution of new technologies. However, there are still some delays and losses in the fundamental process which may be mitigated by deploying alternative technical tools and methods. One such stage is the sensitive pre-drying phase in which ceramic bodies can sustain drying defects such as micro-cracking and fractures due to lack of fine process control. This project investigates the feasibility of using Longwave infrared (LWIR) radiation emitted by a printed Carbon Nanotubes and Graphene (CNTG) heater for pre-drying a clay sample. The CNTG heater emits infrared radiation with a relatively low DC voltage power supply. By radiant heat transfer, homogeneous and uniform drying has been observed in the sample. The penetrative capability of the infrared energy which warms the inside of the sample is presented, as along with the results of comparing the CNTG heater with a silicone mat heater that also emits infrared radiation. The study establishes that the CNTG heater is not only capable of reducing the lead time of ceramics drying using penetrative IR, but also as an efficient and versatile option that can be economically deployed in the pre-drying stage of a ceramic manufacturing process.

Keywords Graphene and carbon nanotubes · Infrared radiation · Ceramics · Drying

Z. Sherif · J. Patsavellas · K. Salonitis (✉)
Sustainable Manufacturing Systems Centre, School of Aerospace, Transport and Manufacturing, Cranfield University, Cranfield, Bedfordshire MK43 0AL, UK
e-mail: k.salonitis@cranfield.ac.uk

Z. Sherif
e-mail: z.sherif@cranfield.ac.uk

J. Patsavellas
e-mail: John.Patsavellas@cranfield.ac.uk

B. Li et al. (eds.), *Advances in Powder and Ceramic Materials Science 2023*, The Minerals, Metals & Materials Series, https://doi.org/10.1007/978-3-031-22622-9_8

Introduction

Ceramic manufacturing is one of the oldest industries known to mankind. In modern times, the production method has been long-established, and the processes involved have adopted technologies and practices that aim to increase productivity whilst minimising waste. Control and monitoring of several processing conditions are taking place in manufacturing plants to ensure the minimisation of process and product waste. This includes the careful handling of the freshly demoulded ceramic products of various shapes, known as green bodies, which inherently carry a high amount of moisture and thus are prone to plastic deformation under even minor pressure [1]. In principle, the moisture content must evaporate steadily and uniformly in order to minimise any differential stresses in the green body that can cause hairline fractures and lead to product defects [2]. However, microclimatic conditions in production plants, such as ambient temperature, airflow, and relative humidity, can be highly irregular and combined with the natural variation of the ceramic materials, frequently resulting in subpar conditions for uniform moisture evaporation during the drying stage [3]. Traditionally, product defects in ceramics plants tend to be high (in comparison with other manufacturing industries) with waste (total loss) accounting for more than 15% of the total yearly production [4].

There is reason to believe that the deployment of low-energy Long-Wave Infrared Radiant (LWIR) to supply highly directional heat to freshly casted ceramic green bodies could be a viable method of effectively and uniformly pre-dry ceramics before entering the main convection-drying part of the manufacturing process. Despite their potential, the utilisation of IR heaters has proven to be largely impractical in many manufacturing plants [5], due to the size and geometrical constraints of the conventional materials such heaters are traditionally fabricated from, comprising bulky and expensive metal structures. Recent advances made in carbon nanotubes and graphene (CNTG) printable inks [6] could enable the development of flexible LWIR heaters due to the novel electric properties of these carbon particles. Such speciality inks can be printed on several different easy-to-fabricate substrates, and using low voltage electricity [7] could offer new IR pre-drying possibilities for ceramics that have until now been costly and/or technically complex.

This research aims to develop and test the feasibility of a heater utilising carbon nanotubes and graphene to supply directional and penetrative Longwave Infrared (LWIR) heat. This should provide energy to pre-dry ceramics by allowing the removal of moisture in a uniform fashion, necessary to achieve 5% of drying shrinkage. This could specifically benefit high value and high volume ceramic manufacturing where the economics of controlled pre-drying is of substantial interest as this process does not require large capital investment or energy, unlike large industrial microwave ovens.

Background

The mechanism of drying a wet ceramic body involves water evaporation. The drying rate can be explained as the rate of moisture loss through evaporation per unit of time. In the manufacturing process, drying is usually carried out in two stages: pre-drying and main drying. Pre-drying is an important phase in ceramic production during which clay plasticity is eliminated and achieves 80–100% of drying shrinkage. Ford [8] states that the drying behaviour of ceramics consists of different phases characterised by milestone points concerning moisture content. At the critical moisture content (CMC) point, further shrinkage of the green ceramic body stops and its drying rate begins to decline. Drying shrinkage occurs when the reduction of water which takes place over an extended period of time results in a reduction in weight. Pre-drying reduces the sudden shrinkage that can lead to the formation of cracks and fractures in the green body from the high temperature of the main dryers. Slow and gradual evaporation of moisture content during pre-drying avoids losses in drying. Shrinkage while drying is significantly one of the most critical factors for the quality control of ceramic products. The contemporary approach to pre-drying observed in a very large ceramic plants involves keeping the cast green bodies for several hours in a holding area (mainly on very slow-moving conveyors) on the shop floor before routing them to an air convection-drying tunnel. This can lead to high irregularity due to the factory's variable ambient environment. Moreover, water evaporation can occur more from the surface of the drying item rather than uniformly from the whole body. This can cause differential shrinking between the exterior and interior wall surface of the ceramic body, generating cross-body tension resulting in micro-fractures that can be detrimental to the quality and as it can lead to total loss after final production [9].

In order to evaporate the water from the ceramic green body at a faster and more uniform rate, heat energy input is required. For instance, the enthalpy of vaporization of water at standard atmospheric pressure is 2260 kJ/kg [10]. Supplying heat to evaporate the water could be achieved by any of the three modes of heat transfer: convection, conduction, and radiation or by a combination of them [8]. The present study focuses particularly on radiative drying of ceramics and offers a more detailed review of the radiant heating characteristics and effects. Penetrative infrared radiation directed to the green body can warm the interior of the solid, and it is generally advantageous as a uniform and efficient heat transfer method. Radiant heat is the emission of electromagnetic waves from any matter that has a temperature greater than absolute zero. Radiant heating technology is defined as the method of intentionally using the principles of electromagnetic radiation to transfer energy from an emitting heat source to another object. Its application can be seen in numerous industries including food, paper, and powder processing [11]. Ford [8] claims that a ceramic that is dried under infrared radiation behaves the same as when dried with convection.

Infrared radiation (IR) is a type of electromagnetic radiation with wavelengths between 700 nm and 1 mm, making it lie between visible light (shorter wavelengths) and microwaves (longer wavelengths) in the electromagnetic spectrum. IR waves are highly directional and travel in straight lines which means that the direction of incident waves upon heating a surface may alter how the radiation reacts with it. Radiation falling on a particular object will get absorbed, reflected, and transmitted. The division of radiation into these three outcomes will mainly depend on the emissivity (ε) of the material. The absorbed radiation is converted to heat energy which will transmit further through the object via conduction and convection [12, 13]. In the instance of using IR to supply heat, radiation is initially absorbed in the surface layer of the object. Subsequently, radiation penetrates to some depth in moist, porous materials. The ability of IR waves to penetrate bodies depends on several factors, including primarily moisture content [14]. This is because water is a great absorber of infrared on account of the presence of O–H bonds, which begin to vibrate at specific frequencies of radiation within the IR part of the electromagnetic spectrum.

The process of converting IR into dipole bond molecule oscillations causes water to evaporate as the kinetic energy imparted in the O–H bonds enables phase change from liquid to vapour [15]. According to Qassem and Kyriacou [16], as the wavelength increases and the spectrum moves deeper into the infrared region, the behaviour of liquid water tends to change remarkably and becomes exceptionally absorbing of radiation. Consequently, the decision to experiment with utilising LWIR is seemingly fitting. According to Woods and Wiedemann [17], the water molecule is Infrared active, meaning that it will bend or stretch causing them to overcome the Van Der Waals forces attracting them together thus enabling water to change phase.

Carbon nanotubes and graphene (CNTG) are known for many novel properties. The main reason for choosing graphene as a radiation element is due to its high emissivity, thermal conductivity, and resistivity. Current literature on heaters made out of CNTG is sparse. Grzegorz Wroblewski et al. [18] used spray-coated multi-walled carbon nanotubes and graphene (MWCNTG) as a transparent heater for a glass panel. The successful spraying proved that nanomaterial coating is a promising and cost-effective approach for fabricating large-area heated glass. In addition, he confirmed that CNTG coatings can be thermally stable and used for varied applications due to their physical and chemical properties including electrical conductivity and radiant heat emissivity. Moreover, according to research conducted by Patsavellas et al. [7], an engineered CNT and Graphene (CNTG) ink mixture was sprayed and cured on the surface of a thin cementitious board (in an area of 50 cm^2) and proved to be an effective electrical LWIR heater emitting radiant heat, reaching 40 °C in 2 min at very low DC voltage (8 V). However, apart from industry specific press, a thorough literature survey revealed a research gap on the topic of CNTG deployment as radiation elements for controlled pre-drying of ceramics. Therefore, a rigorous experimentation method was conducted to ascertain the viability of using CNTG for such purpose. This is presented in the next section.

Methodology

Establishing an effective study of the practical potential of an IR CNTG heater in the specific application required a methodical approach to conducting relevant and rigorous experiments in order to meet the stipulated objectives. Roy and Sutapa [19] state that to study a large number of variables (factors) that impact a given process, a cause-effect relationship should be established that describes how each factor causes a change in response. Using Design of Experiments (DOE) can reveal the possible relationships by establishing function(x) variables = y response. This allows for the factors that cause notable changes in the response (outcomes) of a specific process to be clearly understood by sequential experimentation following a statistical analysis of the results.

Oummadi [1] states that the uncontrolled evaporation rate in clay can lead to defects like cracking and voids in the ceramic body. The objective of conducting the experiments is:

- To control the drying rate during the pre-drying stage of the cast ceramic green body.
- To avoid micro-cracking and voids in the ceramic.
- To minimize the lead time of the pre-drying process.

Patsavellas et al. [7] established that small DC voltages applied to a printed CNTG film material will cause it to emit LWIR heat energy that can penetrate a solid body. Deploying this principle in a suitably designed heater could achieve the desired moisture evaporation at an accelerated but controlled rate while homogeneously pre-drying the green body.

Therefore, the experiments were designed as follows:

1. Sequential experimentation was the chosen method. By conducting a series of experiments, observational data could be collected to establish f(X) = Y.
2. The measurable primary response variables were the loss on drying of the green body and the temperature produced by the heater.
3. The drying rates results were analysed using a regression model which is a powerful statistical tool that establishes the correlation between the control variables and response variables by predicting the appropriate input for the desired output.

The project progressed from focusing on the sole use of a CNTG heater to a comparative approach involving a commercially available silicone heater mat which could act as a radiant heat source and was tested in the same manner as the CNTG heater. Both heaters are shown in Fig. 1. It is worth mentioning that the CNTG heater had its underside painted entirely using a proprietary screen-printing method which deposited a very thin layer along the whole the surface of a Perspex structure using a liquid carrier. Therefore, to achieve a like for like comparison between the two heaters only the top section of the CTNG heater was supplied with voltage and not its sides.

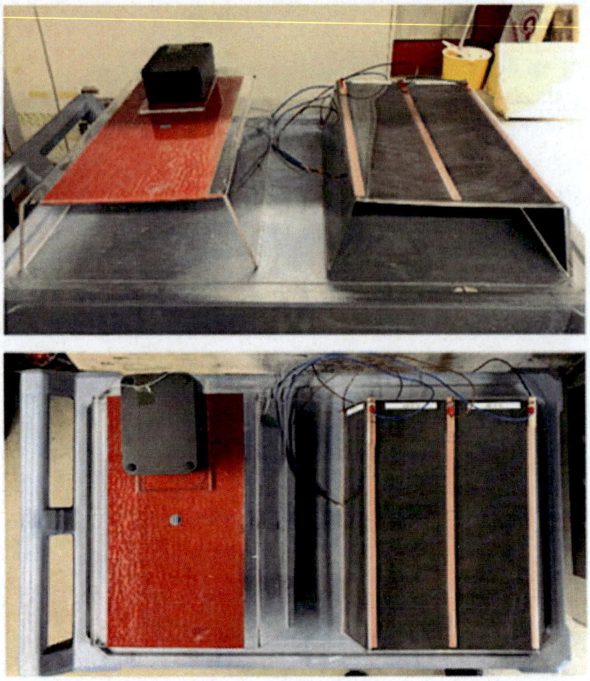

Fig. 1 Silicone heater (Left) and CNTG heater (Right)

The experimental setup was established specifically for testing a variety of radiant heat sources with the purpose of pre-drying ceramic green bodies (Lids) as shown in Fig. 2. According to the capability of conducting this experiment, the heater has been designed as a trapezium-shaped "hood" open at the two long sides in order to allow humidity to escape freely. This hood, fabricated out of clear Perspex material, was sprayed with the specially formulated CNTG ink. The green body ceramic cistern lids were placed on a drying rack as shown in Fig. 3 to reduce surface contact that could block water vapour from escaping from underneath the lid. A digital scale was placed under the lid and the heater to measure the weight loss which was recorded for analysis. Moreover, sensors connected to dataloggers were used to collect temperature and humidity readings. An infrared camera was utilized to capture thermal imaging of the heating process as it occurred to enable the researchers to observe and monitor the homogeneity of the heating as well as the presence of any cold spots.

The data was collected by using a set of devices namely:

- Weighing scale: measuring the weight and calculating the drying rate.
- Stopwatch: To measure elapsed drying time.
- Thermocouple: To measure temperature inside the ceramic sample (error ± 2 °C).
- Voltmeter: To measure applied voltage in the heaters.

Through the experimental setup and procedure, it was identified that the control variables (independent variables) were Voltage and Time. The dependent (response)

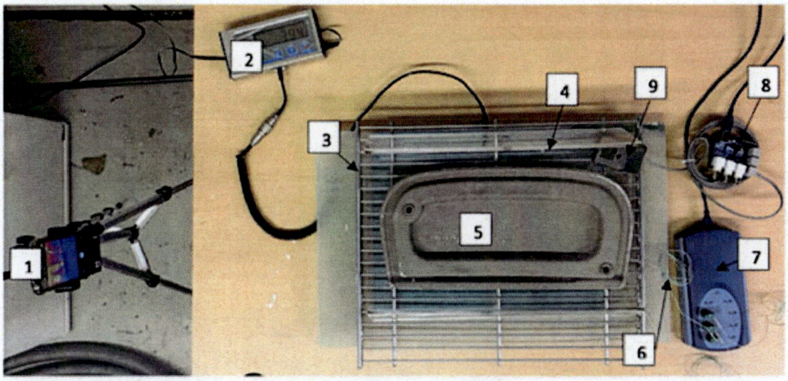

Fig. 2 Experimental setup. (1-infrared camera on a tripod, 2-digital scale, 3-drying rack, 4-heater hood, 5-Lid (green body), 6-K type thermocouples, 7-temperature data logger (TC-08), 8-Humidity Datalogger (Dr Daq), 9-Dr Daq humidity sensor)

Fig. 3 The drying rack

variables were temperature (primary response) and moisture content in the sample (secondary response).

Experimentation

The control variables (X) can be correlated with response variables (Y) as shown in Fig. 4. The voltage supply of the heater and the running time can determine the temperature produced and thereby the drying rate. Initially, a classification experiment was performed with altering voltage and time in 8 trials. These tests enabled the prediction of the levels of both these factors that can result in 5% moisture reduction.

The responses could be measured in a quantitative manner utilising the data collection mechanisms discussed earlier. However, the irregular lab ambient microclimate introduced additional variables which could not be readily controlled. These noise variables (Z) could lead to discrepancies in the measured responses and debase the results. Therefore, an insulating Celotex chamber, shown in Fig. 5, was built to minimise the variation in the environment in which the experiments were being conducted. The other noise factor observed was the lid starting weight which

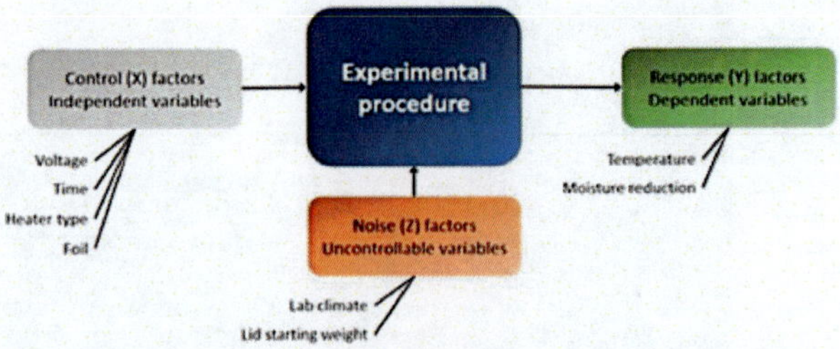

Fig. 4 Relation between response variables and control variables

Fig. 5 Celotex chamber

depended on the casted lid model as well as the starting moisture content, and it is believed that it may have introduced some variations in the results.

Results and Discussion

Firstly, the heater was set to the appointed voltages to allow for measuring the temperatures produced at its surface. It was observed that the temperature is proportional to the voltage supply as seen in Fig. 6. It is important to note that the heater took around 5 min for the heater to reach a stable temperature and hence 5 min were allowed to elapse before a lid was placed underneath it.

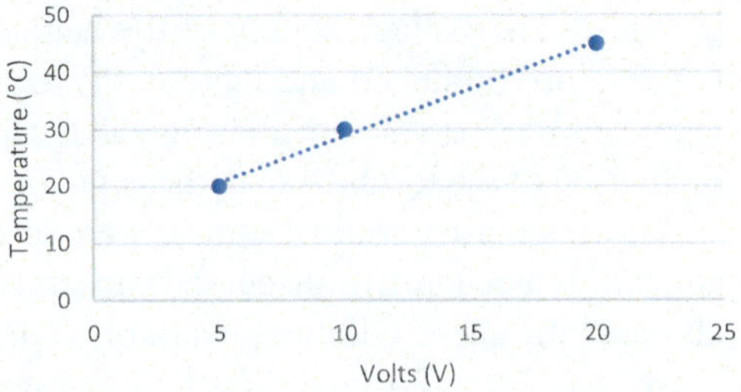

Fig. 6 Supplied voltage and the corresponding temperature

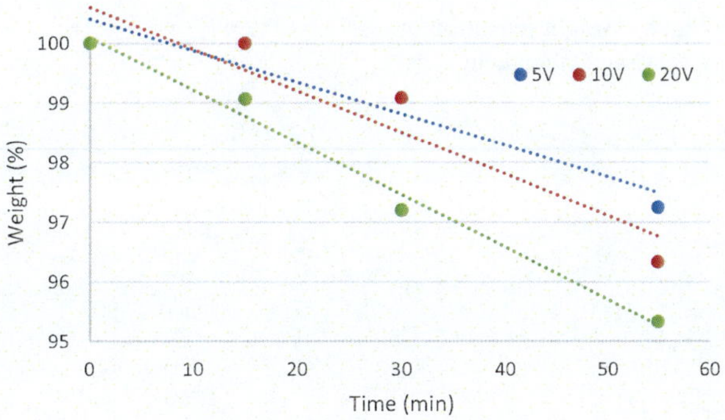

Fig. 7 CNTG classification experiments results

Initial Experiments

After identifying the generated temperatures, green body lids were heated according to the approach presented in Sect. 3. The results are shown in Fig. 7. It can be observed that the weight reduction follows a similar trend for the three applied voltage levels but the rate at which the weight reduces was faster for higher voltages. While 20 V provided the most uniform reduction, 55 min were not enough to achieve the desired 5% reduction. Therefore, the higher the measured temperature, the higher the reduction in weight over the same period. This can be attributed to the Infrared active water molecules [17] which evaporates at a higher rate when exposed to more IR energy.

Comparative and Penetrative Testing

In order to appropriately compare the two heaters, both their surface temperatures were set to 45 °C and the heating was left to take place for 55 min. As seen in Fig. 8, both heaters did not start with the same rate of drying as the silicone heater took longer to reduce the weight of the green body initially. This indicates that the radiation-emitting proficiency of the CNTG heater surpasses that of the silicone heater which could be attributed to higher emissivity.

The energy efficiency of an infrared dryer relates directly to the power required to produce a specific amount of heat [20]. Hence, to further compare the heaters, the power used to produce the same surface temperature was identified by determining the voltage and current supplied to them. The results show that the silicone heater requires 4 times more power to generate matching temperatures (267 W for silicone and 65 for CNTG). This makes the CNTG heater ideal to be used in an industrial setting where minimising the energy running costs is essential.

After demonstrating that utilizing IR can indeed remove moisture from the ceramic green body, its penetrative ability was established by embedding 4 thermocouples in a parallel-line spaced configuration inside the green body where temperature readings

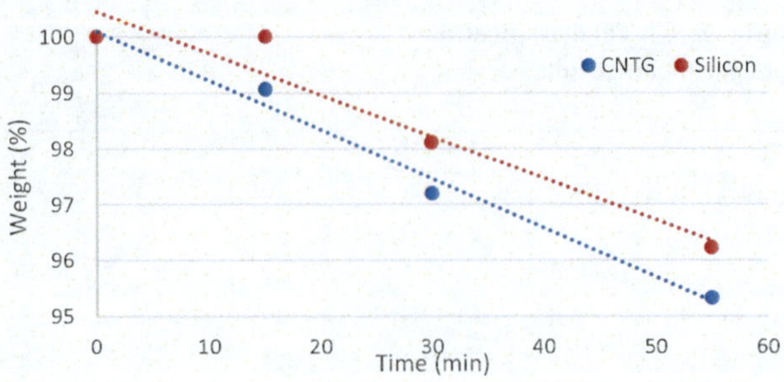

Fig. 8 Weight reduction of CNTG and silicone heater

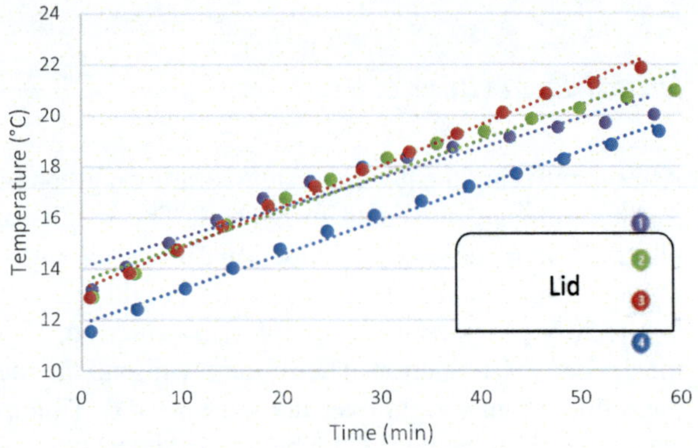

Fig. 9 Buried thermocouples results

were collected every 1 min for an hour. The results are shown in Fig. 9. It is apparent that the energy produced by IR is capable of penetrating through the lid as the temperature rise was both immediate and simultaneous at the 4 points with minor variations.

IR Reflective Testing

As conferred earlier, heat losses may occur due to the reflection of a fraction of the IR rays from the surface of the sample. Therefore, introducing aluminium foil could act as a radiant barrier reducing the heat loss significantly by reflecting the rays back onto the ceramic sample. Consequently, to inspect such attribute, firstly, a rectangular piece of aluminium foil was placed on the non-heating sides of the heater and then on all the heater's outer surface where the CNTG was sprayed, and the results were compared. Figure 10 shows that the weight reduction in both configurations was similar for the first 15 min. However, as time passed, some changes started to appear on the heater with foil only in the sides, which showed a significant reduction of 5.46% compared to 3.81% with the foil placed all around. While the result proves that radiant heat could be reflected and directed, having the foil on the heating section

Fig. 10 IR reflective testing with aluminium foil

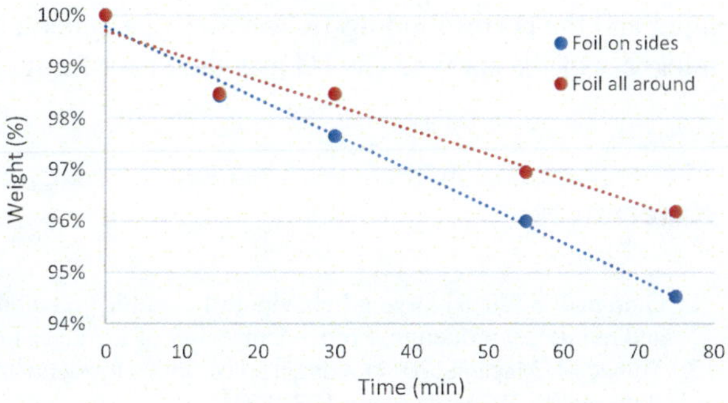

of the heater may have caused some heat dissipation via conduction and therefore less heat is reaching the sample.

Conclusion

The project was established to examine the viability of utilising CNTG as an emitter of LWIR for pre-drying ceramics and the possibility of deploying it in a manufacturing setup. The science concerning the behaviour of ceramics while drying and the principal characteristics that influence IR emittance were investigated to aid the justification of using CNTG. The experiments conducted in the research allowed for the recognition of the CNTG heater's ability to not only dry the ceramic green bodies in a uniform and homogeneous manner but also reduce the time needed for the process. This was presented by successfully reaching 5% reduction in weight of the ceramic green body. Moreover, it provided an insight into the radiant emittance and power efficiency of CNTG in such application. The ability to spray paint CNTG on any surface makes it highly versatile and therefore could be utilised to produce custom heaters of various shapes and sizes to irradiate every section of the complex contours of different ceramics products.

Cracking is one of the most common defects in ceramics production [21]. Consequently, reducing production losses that are caused by cracking at the pre-drying stage can lead to substantial financial savings as it will decrease the amount of rejected products that would need to be recycled or remanufactured. Therefore, a project to industrialise the findings may be compelling. On that account, the experiments could be repeated by building on the findings with 4 control variables full factorial experimental design in a randomised fashion at the factory floor. Moreover, as expressed in the literature review, 6 μm is one of the wavelengths at which the water molecule bends resulting in the release of the Van Der Waals forces and allowing water to change phase from liquid to gas. According to Winn's displacement law, it was estimated that the CNTG heater produced IR with a peak wavelength of 9 μm. However, it would be immensely beneficial to be able to accurately identify the wavelength produced using a spectrometer. Further work could aim to classify the varying power input and the corresponding IR wavelengths generated as well as including varying thickness of the sprayed CNTG as a factor for heater design optimization.

References

1. Oummadi S (2019) Drying behaviour of ceramic green bodies: experimental characterization and numerical modeling. France University of Limoges Ph.D. thesis
2. Trunec M, Maca K (2014) Advanced ceramics for dentistry. In: Advanced ceramic processes. Butterworth-Heinemann, pp 123–150
3. Zaccaron A, Nandi SdV, Bernardin AM (2021) Fast drying for the manufacturing of clay ceramics using natural clays. J Build Eng 33
4. Suryadi A, Ardiansyah F, Ngatilah Y (2018) Quality analysis of ceramic tent product with six sigma method in PT. Mas Keramik KIA. In: The 2nd international joint conference on science and technology (IJCST)
5. Agrafiotis C, Tsoutsos T (2001) Energy saving technologies in the European ceramic sector: a systematic review. Appl Therm Eng 21(12):1231–1249
6. Nguyen BH, Nguyen VH (2016) Promising applications of graphene and graphene-based nanostructures. In: Advances in natural sciences: nanoscience and nanotechnology (ANSN), vol 7, p 023002
7. Patsavellas J, Salonitis K, Koziol K, Zakrzewski L, Blackwood B (2020) Carbon nanotubes and graphene radiant heater printed on a cementitious flooring substrate: a feasibility study
8. Ford RW (1986) Ceramic drying. Pergamon Press
9. Robinson DA, Brosnan GC (2003) Introduction to drying of ceramics with laboratory exercises. The American Ceramic Society, Ohio
10. Datt P (2014) Latent heat of vaporization/condensation. In: Singh VP, Singh P, Haritashya UK (eds) Encyclopedia of snow, ice and glaciers. Encyclopedia of earth sciences series. Springer, Dordrecht
11. Ranjan R, Irudayaraj J, Jun S (2002) Simulation of infrared drying process. Drying Technol 20(2):363–379
12. Riadh MH, Ahmad SAB, Marhaban MH, Soh AC (2015) Infrared heating in food drying: an overview. Drying Technol 33(3):322–335
13. Noboru S, Tamotsu H (1994) Applications and advances in far-infrared heating in Japan. Trends Food Sci Technol 5(11):357–362
14. Abukhalifeh H, Dhib R, Fayed ME (2005) Model predictive control of an infrared-convective dryer. Drying Technol 23(3):497–511
15. Aboud SA, Altemimi AB, Al-HiIphy ARS, Yi-Chen L, Cacciola F (2019) A comprehensive review on infrared heating applications in food processing. Molecules 24(22)
16. Qassem M, Kyriacou PA (2013) Comparing the rates of absorption and weight loss during a desorption test using near infrared spectroscopy. Skin Res Technol 19(2)
17. Woods KN, Wiedemann H (2004) The relationship between dynamics and structure in the farinfrared absorption spectrum of liquid water. Chem Phys Lett 393(1–3):159–165
18. Wroblewski G, Sloma M, Janczak D, Sarapuk L-D, Jakubowska M (2016) Large area, transparent heaters based on carbon nanotubes and graphene platelets for heated glass application. In: 20th European microelectronics and packaging conference and exhibition: enabling technologies for a better life and future
19. Roy SK, Sutapa IN (2003) Case studies of use of design of experiments in material research in Fajultas Teknologi industri 5(1):32–44
20. Pawar SB, Thorat BN (2011) Infrared drying of alumina-silicate mineral cake. Drying Technol 29(7):819–824
21. Lakhdar Y, Tuck C, Binner J, Terry A, Goodridge R (2021) Additive manufacturing of advanced ceramic materials. Progress Mater Sci 116

Enhancing Reinforcing Efficiency of SiC Particles in Aluminum Matrix Composites with Intercalated Oxygen Atoms

M. R. Joo and D. H. Bae

Abstract In metal matrix composites (MMCs), the interface between the metal matrix and reinforcement critically influences mechanical properties of MMCs because the load can transfer at the interface via the interfacial shear stress. We develop a new aluminum matrix composite (Al-O/SiC composite) reinforced with silicon carbide (SiC) particles which are dispersed in an interstitial aluminum alloy (called as I-Al) matrix containing oxygen atoms. The new composites reinforced with SiC particles are fabricated via powder metallurgy. Mechanical milling induces uniform dispersion of SiC particles in the matrix and also enables oxygen in the I-Al to be redistributed in the composite powder. Oxygen is observed at the interface between the matrix and SiC particles in the Al-O/SiC composite. Oxygen at the interface can provide additional chemical bonding other than mechanical interlocking at the interface, improving interface bonding of the composites. With the same amount of SiC particles, the mechanical properties of the Al-O/SiC composites including Vickers hardness, compressive yield strength, and elastic modulus are enhanced as compared to those of the Al/SiC composites. Therefore, the intercalated oxygen contributes to increasing reinforcing efficiency of the Al-O/SiC composites.

Keywords Aluminum matrix composites · Interfaces · Mechanical properties · Reinforcing efficiency

Introduction

Aluminum matrix composites (AMCs) reinforced with ceramic particles have received great interest in automobile and aerospace industries for several decades due to high specific strength and elastic modulus, and superior creep and wear resistance [1]. Extensive research has been carried out to obtain high-performance AMCs with excellent mechanical properties. Especially, the AMCs containing particulate reinforcements show isotropic mechanical properties, and their fabrication processes

M. R. Joo · D. H. Bae (✉)
Department of Materials Science and Engineering, Yonsei University, Seoul 03722, Republic of Korea
e-mail: donghyun@yonsei.ac.kr

B. Li et al. (eds.), *Advances in Powder and Ceramic Materials Science 2023*, The Minerals, Metals & Materials Series, https://doi.org/10.1007/978-3-031-22622-9_9

are relatively easier and cheaper as compared with the AMCs reinforced with fiber-type ones [2]. The AMCs with various particulate reinforcements including silicon carbide (SiC), boron carbide (B_4C), titanium oxide (TiO_2), and aluminum oxide (Al_2O_3) achieve excellent mechanical properties. Especially, SiC particles exhibit low density, high elastic modulus, and specific strength [3] as well as excellent wear resistance and good thermal conductivity, which is a superior candidate to reinforce aluminum matrix.

In metal matrix composites including AMCs, the interface between the metal matrix and reinforcements plays an important role in mechanical properties [4]. The main strengthening mechanism of MMCs is a load transfer through the interface when interfacial shear stress is applied. The interface is closely related to local environments around the interface including interfacial structure, interfacial bonding, an alignment of reinforcements in the matrix [5], and so on. Thus, the characteristics of interfaces significantly influence mechanical properties in MMCs, so it should be noted that developing interfaces is a critical issue to achieve efficient load transfer from the matrix to the reinforcements when MMCs are fabricated.

MMCs have been fabricated by using the liquid-state process and solid-state process. However, when utilizing the liquid-state process, it is difficult to uniformly disperse reinforcements in the metal matrix. Also, unavoidable by-products are formed easily in the matrix due to high-temperature process, which can result in deteriorating mechanical properties of MMCs. Powder metallurgy (PM) technique is one of the most popular processes in solid-state process to fabricate AMCs. Mechanical milling is the most common process to fabricate the composite powder, which enable uniform dispersion of reinforcements. The interface characteristics are largely determined by optimizing processing parameters during mechanical milling. When fabricating AMCs reinforced with particulate reinforcements, it is important to uniformly disperse particles in the aluminum matrix. Individually dispersed particles in the matrix offer a large interface area to exploit high load transfer efficiency, which strongly determines reinforcing efficiency of the composites.

In this study, new AMCs where SiC particles reinforce the aluminum matrix (I-Al) containing oxygen atoms are fabricated via powder metallurgy. By using mechanical milling, SiC particles are uniformly dispersed in the I-Al, followed by consolidation process using hot-pressing. Interface characteristics of the composites are investigated. Furthermore, mechanical properties including Vickers hardness, compressive strength, and elastic modulus of the composites are evaluated.

Experimental

Pure aluminum powder (~150 μm in average and 99% purity) is supplied from Kojundo Korea CO., LTD. SiC particles with a size of ~1 μm in average (99.99% purity, Avention) and SiC nanoparticles with a size of 60 ~80 nm in average (99% purity, Ditto technology) are prepared as a reinforcement. The new aluminum powder (Al-O powder) was produced using a gas atomizer. Before gas atomization, aluminum ingots with 99.8% purity were melted in a crucible in a melting furnace at 720 °C using the conventional gravity casting method. Oxide nanoparticles with a size of ~20 nm in average wrapped in aluminum foil were inserted and stirred in the melt. The melt was poured into rectangular stainless-steel molds preheated to 100 °C. Then, gas atomization of Al-O powder was carried out with ultra-high-purity argon atomization gas and a discrete jet, high-pressure gas atomization nozzle with a pouring temperature 900 °C. The gas-atomized aluminum powder and conventional aluminum powder with 5 vol.% SiC particles were mechanically milled with stainless steel balls (~5 mm in diameter) using an attrition milling at 500 rpm. During mechanical milling, the ball-to-powder ratio was set to 15:1, and circulating water was used to minimize temperature increase. Stearic acid (1 wt.%) was used as a process control agent to prevent excessive cold welding of the powder. The milled powders were consolidated by using hot-pressing process (Carver, Inc., USA) at a temperature of 500 °C for 1 h with an applied pressure of ~270 MPa.

The microstructure of powder specimens was investigated by scanning electron microscopy (SEM; JSM-7001F, JEOL, Japan), wherein the composition at the surface of the composite powder was analyzed by energy dispersive spectrometer (EDS; Octane plus, AMETEK, Inc., USA). The cross-sectional microstructure of composite powder was carried out using optical microscopy (OM; Eclipse LV150, Nikon, Japan). The specimens were prepared by mechanical polishing until they became the mirror finish. To investigate the consolidated composites, transmission electron microscopy (TEM; JEM-ARM200F, JEOL, Japan) was utilized. By using an ion beam milling method (PIPS 691; Gatan, Oxford, U.K.), thin foil specimens which were used in TEM analysis were prepared. The hardness of the specimens was measured by using a micro-Vickers hardness testing machine (MXT-α76, Matsuzawa Seiki Co., LTD., Japan) with an indenter load of 300 gf. Before hardness test, the specimens were mechanically polished using SiC papers up to #2000. The uniaxial compressive tests were conducted under a constant cross-head speed condition using an instron-type machine (RB301, R&B, Korea) with an initial strain rate of 1×10^{-4} s^{-1} at room temperature. The specimens were mechanically polished, and the shape of them was rectangular with a 2:1 height-to-width ratio. During compressive tests, two tungsten carbide plates were placed on the top and bottom of the specimen and sprayed by BN to minimize the effect of friction. The elastic modulus of composites was measured by an ultrasonic measuring system (Hankooklab, HKL-01-UEMT, Korea) and calculated by pulse-echo method.

Results and Discussion

Figure 1 shows SEM images of gas-atomized Al-O powder (a-b) and as-received SiC particles (c-d). Al-O powder was produced via gas atomization of which the process includes remelting an aluminum alloy (I-Al) containing oxygen atoms. Gas-atomized Al-O powder has a size of ~54 μm in average which is calculated by measuring diameters of powder in SEM images. As-received SiC particles seem like

Fig. 1 SEM images of gas-atomized Al-O powder (**a**, **b**) and as-received SiC particles (**c**, **d**)

a size of less than 100 μm as shown in Fig. 1c. The voids among small SiC particles with a size of ~1 μm are observed at a surface of a SiC powder in a magnified image of a SiC powder in Fig. 1d. Sub-micron SiC particles are aggregated and agglomerated, so the size of aggregated SiC particles seems to be over several tens of micrometers in Fig. 1c. Before gas-atomizing the I-Al, when fabricating the I-Al [6], loosely agglomerated oxide nanoparticles are intentionally soaked in the aluminum melt. The nanoparticles can be decomposed in the melt [7] due to the high chemical potential energy originating from the surface curvature of the extremely small nanoparticles [8, 9]. However, oxygen atoms are rarely soluble in the aluminum melt, the oxygen and aluminum atoms from Al-O clusters, which are observed in a form of nanofibers after solidification [6]. Thus, oxygen atoms in the I-Al are also contained in the gas-atomized Al-O powder during the gas atomization process.

Mechanical milling is known as one of the advantageous techniques to fabricate metal matrix composite due to uniform dispersion of reinforcements in a metal matrix. So, mechanical milling is used to homogeneously disperse SiC particles in the aluminum matrix in this study. Figure 2 shows SEM images of mechanically milled Al/SiC composite powder. The morphology of Al/SiC composite powder is different from that of as-received Al powder due to plastic deformation of composite powder during mechanical milling. As shown in a magnified image of Al/SiC composite powder in the Fig. 2b, the morphology of the Al/SiC composite powder seems to be irregular as compared with that of the as-received one.

Fig. 2 SEM images of mechanically milled Al/SiC composite powder

During mechanical milling, kinetic energy of milling media induces plastic deformation of powders. High ratio of ball to powder and high rotation speed contribute to sufficient plastic deformation of aluminum powder. Relatively soft aluminum powder becomes shattered and flattened due to collisions with milling media and hard SiC particles during milling process. In the meantime, both aluminum powder and SiC particles receive compressive forces from collisions [10], producing plastic deformation of aluminum and SiC particles.

Cross-sectional optical microscope (OM) images of mechanically milled Al-O/SiC composite powder are shown in Fig. 3. The Al-O/SiC composite powder in Fig. 3a has a size of ~200 μm. Many dark gray particles with sub-microsizes are observed inside the composite particle in a magnified image of the composite particle as shown in Fig. 3b. It seems that the aggregated SiC particles over several micrometers are fragmented and the fragmented SiC particles are homogeneously distributed in the matrix during mechanical milling. Many SiC particles are observed in the aluminum matrix according to the cross-sectional OM images. Some shattered SiC particles are observed at a surface of the composite powder. Fig. 4a shows a surface of the composite powder. Region #1 and #2 in Fig. 4a represent the matrix and the SiC particle. Figure 4b and c indicate elemental analysis at region #1 and #2, respectively. The concentration of each composition in region #1 and #2 is summarized in Table 1. Region #1 mostly consists of aluminum, and Region #2 contains silicon and carbon as well as aluminum, indicating the existence of a SiC particle. Carbon concentration provided in the Table 1 is relatively low to identify SiC particles. However, we can still identify SiC particles in the composite based on a sufficient silicon concentration originating from SiC particles. Also, carbon that is detected in the composite can only originate from SiC particles in the composite, of which the carbon concentration (~2.03 wt.%) is higher than that (~0.13 wt.%) in the matrix. Although SiC particles are observed at a surface of the composite powder, most SiC particles are embedded into the aluminum matrix. Thus, SiC particles are homogeneously dispersed in the matrix.

Mechanical properties of the Al/SiC and the Al-O/SiC composites are investigated by Vickers hardness and compressive tests as shown in Fig. 5a and b, respectively.

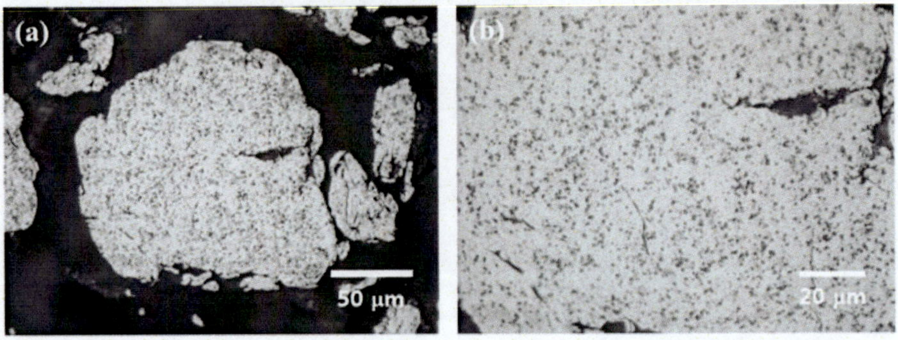

Fig. 3 Cross-sectional OM images of mechanically milled Al-O/SiC composite powder

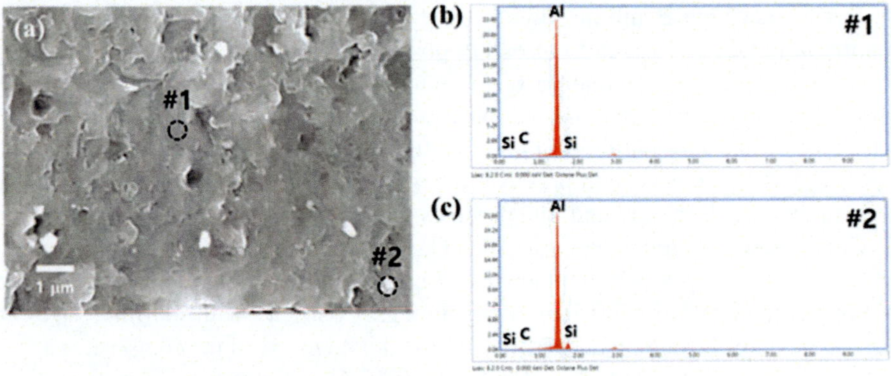

Fig. 4 A SEM image of a surface of Al-O/SiC composite powder in **a**, and EDS data of aluminum matrix and a SiC particle (**b**, **c**)

Table 1 An elemental analysis of Si, C, and Al according to the EDS data of region #1 and #2 in Fig. 4

wt.%	Si	C	Al
Region #1	0.02	0.13	99.85
Region #2	8.74	2.03	89.23

The Vickers hardness of Al-O/SiC composites has a higher value of ~151.2 Hv than that of Al/SiC composites ~143.5 Hv. The compressive strength of both composites corresponds to the values of Vickers hardness. The compressive yield strength of Al-O/SiC composites is ~401.9 MPa, which is higher than that of Al/SiC composites, ~356.7 MPa. Furthermore, the elastic modulus of Al/SiC and Al-O/SiC composites is studied. The elastic modulus of Al/SiC and Al-O/SiC composites is ~77.2 GPa and ~80.0 GPa, respectively, as shown in Fig. 6.

Figure 7 shows TEM images of the Al-O/SiC composites and line scan data across the SiC particle in Al-O/SiC composites. A SiC particle with a size of ~200 nm is observed in the aluminum matrix as shown in Fig. 7a. A magnified image of the

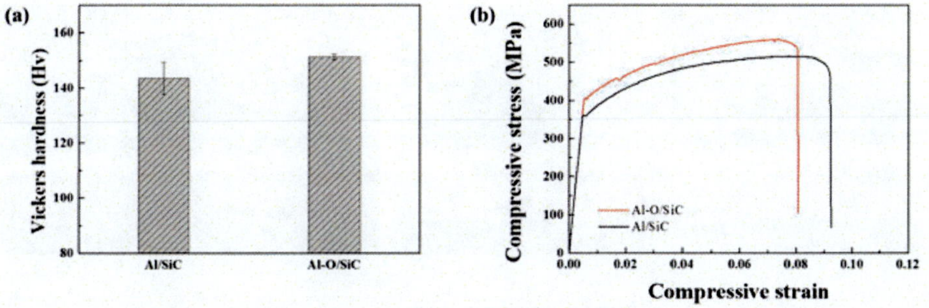

Fig. 5 Vickers hardness and compressive stress-strain curves of Al/SiC and Al-O/SiC composites

Fig. 6 Elastic modulus of Al/SiC and Al-O/SiC composites

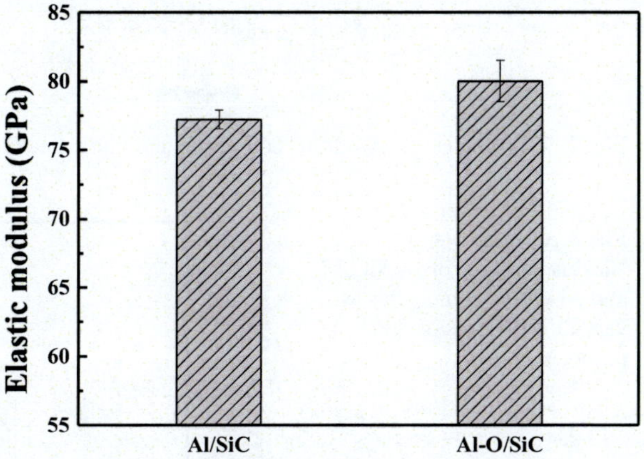

SiC particles in Fig. 7a is shown in Fig. 7b. The SiC particle has an irregular shape since it might be attributed that the particles receive compressive stress when it is detached from aggregated SiC particles and fragmented during mechanical milling. The composition profile of Si, C, and O, determined by using EDS analysis, along the red line marked in Fig. 7b shows that oxygen atoms are detected at the interface (marked by orange dash lines in Fig. 7c) and the concentration of oxygen at the interface is relatively higher than that in adjacent regions. To investigate the interface structure in the Al-O/SiC composites, the fast Fourier transform (FFT) patterns of region #1 and #2 in Fig. 7d are compared. Region #1 represents the aluminum matrix according to the FFT pattern of region #1. The FFT pattern of region #2 shows two types of structures including SiC structure and the aluminum matrix. Other than two types of structure, other compounds related to oxygen are not observed at the interface as shown in Fig. 7d. It suggests that oxygen atoms at the interface are rarely consumed for oxide formation at the interface, so they might be intercalated at the interface between the aluminum matrix and SiC particles.

Comparison of Vickers hardness of the Al/SiC and the Al-O/SiC composites reinforced with different sizes of SiC particles is presented in Fig. 8. Al/SiC-m

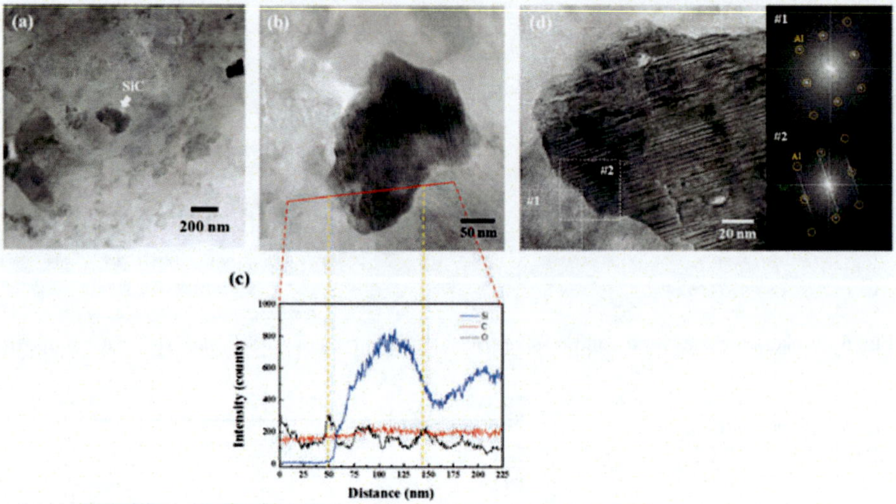

Fig. 7 TEM images of Al-O/SiC composite and line scan data of Al-O/SiC composite

Fig. 8 Comparison of Vickers hardness of Al/SiC and Al-O/SiC composites as varying with sizes of SiC particles

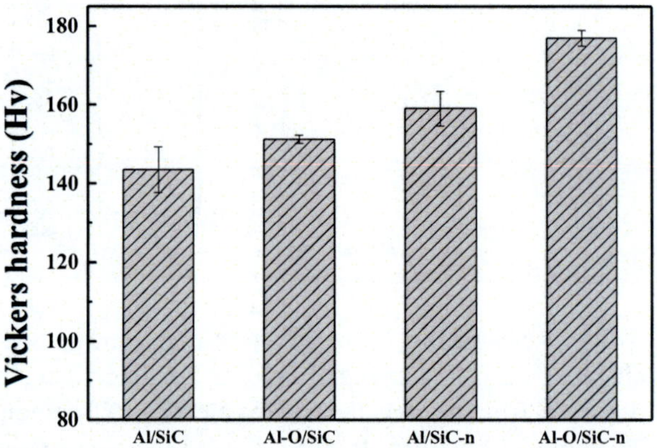

and Al-O/SiC-m composites represent the Al/SiC and the Al-O/SiC composites reinforced with microsized SiC particles, respectively. Likewise, Al/SiC-n and Al-O/SiC-n composites represent the Al/SiC and the Al-O/SiC composites reinforced with nanosized SiC particles, respectively. As mentioned in Fig. 5a, Al-O/SiC-m composites have a higher value of Vickers hardness than Al/SiC-m composites since it might be attributed to intercalated oxygen layers at the interface in Al-O/SiC-m composites. Non-transition metals such as aluminum lack *d*-sub shell and have very limited affinity with ceramic particles [5], providing mechanical interlocking with ceramic particles in the AMCs rather than forming ionic bonds. Intercalated oxygen atoms at the interface in the Al-O/SiC composites can provide the potential for additional chemical bonding at the interface by bridging the matrix and SiC particles. The Vickers hardness of Al/SiC-n and Al-O/SiC-n composites shows a similar trend that the Vickers hardness of Al-O/SiC-n composites is higher than that of Al/SiC-n composites. However, Al-O/SiC-n composites exhibit a higher value of Vickers hardness of ~176.9 Hv as compared to Al-O/SiC-m composites due to the size effect of SiC particles. SiC nanoparticles provide a larger interfacial area as compared to microsized ones, enhancing reinforcing efficiency of SiC in the Al-O/SiC-n composites.

Conclusion

In this study, the new AMCs (Al-O/SiC composites) where SiC particles reinforce the I-Al matrix containing oxygen atoms are fabricated via powder metallurgy. Mechanical milling induces uniform dispersion of SiC particles in the matrix, followed by hot-press consolidation process. The Vickers hardness and compressive strength of the Al-O/SiC composites are improved as compared with those of the Al/SiC composites. Furthermore, the Al-O/SiC composites show a higher elastic modulus than the Al/SiC composites. It might be attributed to the intercalated oxygen layers which are observed at the interface in the Al-O/SiC composites. The layers can lead to forming additional interface bonding rather than mechanical interlocking that is also formed at the interface in the Al/SiC composites. Thus, this new strategy for fabricating AMCs opens a new pathway for achieving excellent performances, which would be of significant importance to the relevant researchers.

References

1. Kamrani S, Simchi A, Riedel R, Seyed Reihani SM (2007) Effect of reinforcement volume fraction on mechanical alloying of Al-SiC nanocomposite powders. Powder Metall 50:276–282. https://doi.org/10.1179/174329007X189621
2. Miracle DB (2005) Metal matrix composites - from science to technological significance. Compos Sci Technol 65:2526–2540. https://doi.org/10.1016/j.compscitech.2005.05.027
3. Baron C, Springer H (2017) Properties of particle phases for metal-matrix-composite design. Data Br 12:692–708. https://doi.org/10.1016/j.dib.2017.04.038
4. Bakshi SR, Lahiri D (2010) A Agarwal, CARBON nanotube reinforced metal matrix composites - a review. Int Mater Rev 55:41–64. https://doi.org/10.1179/095066009X12572530170543
5. Shin SE, Choi HJ, Hwang JY, Bae DH (2015) Strengthening behavior of carbon/metal nanocomposites. Sci Rep 5:16114. https://doi.org/10.1016/j.carbon.2014.10.044
6. Jeon JH, Jeon JG, Joo MR, Lee JW, Bae DH (2022) Deformation behavior of an A356 alloy containing small sub-grains with wide low-angle boundary. J Alloys Compd 908:164550. https://doi.org/10.1016/j.jallcom.2022.164550
7. Shin JH, Choi HJ, Cho MK, Bae DH (2014) Effect of the interface layer on the mechanical behavior of TiO2 nanoparticle reinforced aluminum matrix composites. J Compos Mater 48:99–106. https://doi.org/10.1177/0021998312469238
8. Levitas VI, Pantoya ML, Chauhan G, Rivero I (2009) Effect of the alumina shell on the melting temperature depression for aluminum nanoparticles. J Phys Chem C 113:14088–14096

9. Levitas VI, Samani K (2011) Size and mechanics effects in surface-induced melting of nanoparticles. Nat Commun 2:284. https://doi.org/10.1038/ncomms1275
10. Tarantili P, Andreopoulos A, Galiotis C (1998) Real-time micro-Raman measurements on stressed polyethylene fibers. 1. Strain rate effects and molecular stress redistribution. Macromolecules 31:6964–6976. http://pubs.acs.org/doi/abs/https://doi.org/10.1021/ma961498l

Weathering Resistance of Post-consumer Glass and Sawdust Reinforced Polyester Composites

Kator Jeff Jomboh, Mohammed Kabir Yakubu, Wilson Uzochukwu Eze, Adele Dzikwi Garkida, and Emmanuel Majiyebo Alemaka

Abstract Weathering resistance test was carried out to determine the ability of post-consumer glass and sawdust reinforced polyester composite to withstand outdoor service conditions using water absorption (WA) and tensile properties. 16 samples were investigated consisting of various percentage by weight of post-consumer glass, sawdust as well as hybrid compositions. Test pieces were submerged in distilled water for 35 days and weighed on a daily basis to determine the level of absorption. Changes in tensile strength before and after water absorption were recorded and compared. Samples 5/27.5/67.5, 0/40/60, and 0/20/80 wt.% of sawdust/post-consumer glass/polyester had the lowest WA of 1.23%, 1.91%, 1.36% respectively while hybrid sample 20/20/60 wt.% had the highest WA of 7.46%. Hybrid samples 5/27.5/67.5, 7/33/60, 20/20/60 wt.% with original tensile strength of 11.60, 19.76, 10.25 MPa respectively had an improved tensile strength of 12.63, 22.84, and 12.85 MPa respectively. Generally, tensile strength increased after water absorption indicating increase in weathering resistance which suggests that the composite material can be employed as particle board for outdoor application.

Keywords Weathering resistance · Tensile strength · Water absorption · Particle board

K. J. Jomboh (✉)
Department of Industrial Design, University of Maiduguri, Borno State, Maiduguri, Nigeria
e-mail: efferyjom@yahoo.com

M. K. Yakubu
Department of Polymer and Textile Engineering, Ahmadu Bello University, Zaria, Nigeria

W. U. Eze
Department of Polymer Technology, Nigerian Institute of Leather and Science Technology, Zaria, Nigeria

A. D. Garkida · E. M. Alemaka
Department of Glass and Silicate Technology, Ahmadu Bello University, Zaria, Nigeria

B. Li et al. (eds.), *Advances in Powder and Ceramic Materials Science 2023*, The Minerals, Metals & Materials Series, https://doi.org/10.1007/978-3-031-22622-9_10

Introduction

Composites as valuable and versatile family of engineered materials are used to solve problems of different applications, improve productivity, lower cost, and facilitate the introduction of new properties in materials [1]. The characteristic properties of these composites are as a result of the individual properties of their constituent fragments and their respective volume fractions and arrangements in the material system [2]. Polymer matrix composites such as polyesters are known to display ideal matrix materials due to their low cost, low density, ease of processing as well as desirable mechanical features. They are either thermosets or thermoplastics polymers [3]. Over time, the polymer matrix composites have found applications in the building and construction industry either as; panels for partition and false ceiling, partition boards, wall, floor, window and door frames, roof tiles, and among others [4]. The quest for high-quality, locally sourced economic, and eco-friendly building materials has led to on ending assessment on the viability of wastes for particle board production in the past decade [5]. Waste materials with both thermoplastics and thermosetting matrices have been used to develop various engineered composites boards. Among efforts to improve the properties of particle board composites, an invention involving the conversion of waste plastics coating powder and waste paint was combined with various fillers to make composite boards of varying properties such as densities. Examples include waterproof composite boards made from sawdust with potential usage where normal medium density fibreboard "MDF" is not suitable [6]. The development of high-performance particleboards from a mixture of wood particles and short glass fibres using a vacuum-assisted resin transfer technique revealed excellent mechanical properties, water resistance, and dimensional stability as compared to commercial wood composites [7]. Improved tensile and impact strength have also been reported in sisal fibre and silicon carbide filler reinforced composites [8].

Weathering test is a tool used in decision making to determine how a material responds to certain weather conditions and can be used to anticipate potential performance problems [9, 10]. The accelerated test is carried out due to lack of time, waiting for years to be sure if a product will perform adequately in an intended end-use environment. It may be used to test if material formulations can maintain its properties by resisting degradation and colour change when exposed to temperature changes such as water absorption and ultraviolet UV radiation [11]. For this study, weathering resistance was measured to ensure performance ability of post-consumer glass and sawdust reinforced polyester composites to withstand outdoor service conditions using water absorption (WA) and tensile properties. Changes in tensile strength properties before and after water absorption were evaluated.

The water immersion test is usually used to study the adsorption behaviour of materials from which the water diffusion mechanisms of the composites are established. A study of polyester matrix composites reinforced with oil palm ash and oil palm fibre reported that the amount of water absorbed by the composites increases with increase in the fibre weight ratio of the hybrid reinforcement and also with the increase in the weight percent of the reinforcement phase [12]. Hassan and Awopetu

[13] in the production of particle boards from sawdust, corncobs, and rice husk, using polyester resin reported water absorption range of between 5.44% to 47.77% with 5.44% as the optimum absorption rate. Studies have shown absorption as one of the major disadvantages associated to natural fibre composites which also reduces mechanical performance with the exception of impact energy which is commonly seen to increase [14]. In other to achieve improved or superior performance properties in composites such as moisture resistance, hybridization of natural and synthetic fibres is encouraged [15, 16]. Kim [17] opined that by hybridization of natural and synthetic fibres, it is possible to achieve a balance between performance properties and cost of the composites, which would not otherwise be obtained with a single kind of reinforcement. For this study, post-consumer glass and sawdust hybrid reinforcement was used in a polyester matrix to improve the weathering resistance properties of the composites.

Methodology

The thermoset polymer matrix composite was utilized for this experiment with post-consumer glass and sawdust in particulate form as reinforcement while unsaturated polyester resin was used as the matrix to develop composites particle board by resin casting technique using a D-optimal mixture experiment. Post-consumer glass sourced within Samaru-Zaria was sorted off heavy impurities, washed, and dried. Size reduction was done using a metal hammer before more crushing was done in an electric crushing machine. Sawdust gotten from a sawmill was dried in a Cole-Parmer vacuum oven at a temperature of 90 °C for 24 h. The crushed cullet and sawdust aggregates were both sieved and the particle sizes of 0.5–0.24 mm determined. The composites were then made using the different mixture components of post-consumer glass, sawdust, and polyester resin transferred in an open glass mould of 150 × 200 × 5mm and allowed to cure and set for 24 h. The mould was covered with aluminium sheet with paraffin gel to serve as releasing agent to enable free release of the cured composite. Weathering characteristics were investigated to predict the durability of the composites and to monitor changes in tensile strength properties. Tensile test samples were machined, cleaned, and submerged in distilled water for 35 days (see Fig. 2). They were then removed, dried, and weighed on a daily basis using a digital weighing balance to determine the absorption level of each sample material. The procedure was repeated until there was no obvious increase in weight of samples in accordance to ASTM D5229. Weight change was then expressed in percent using the formula; Percent Weight Change = Final weight − Initial weight/Initial weight × 100. Thereafter, tensile test was done on the wet samples using a universal tensile machine in accordance to ASTM D638 standard. Dumbbell shaped specimen sizes of 100 × 15 × 5 mm were placed in the grips of the machine and pulled until failure (see Figs. 1 and 2). A crosshead speed of 2 mm/min was used with a maximum force of 10KN. Tensile strength (MPa), modulus of elasticity (MPA), and elongation (%) were calculated from the test results. Tensile test was conducted on both dry samples

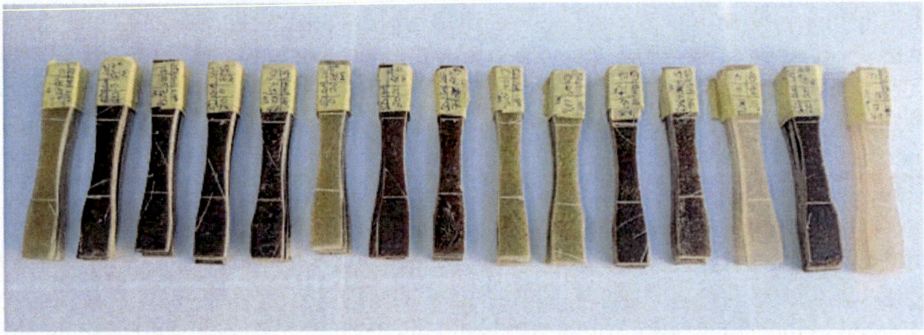

Fig. 1 Tensile test samples of post-consumer glass and sawdust reinforced polyester composites

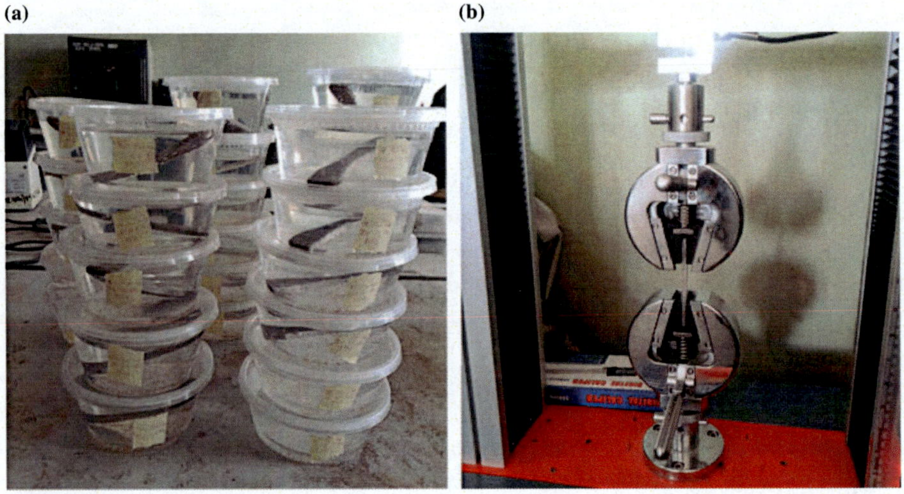

Fig. 2 (**a**) and (**b**) Weathering resistance procedure showing water absorption setup and tensile strength analysis

and wet samples (samples that underwent weathering through water absorption) and then compared.

Results and Discussion

Water Absorption

Water absorption result as presented in Fig. 3 shows an increase in absorption with increase in immersion time for a period ranging from 20 to 24 days before saturation point. Also, water absorption increased along with increase in filler loading with the

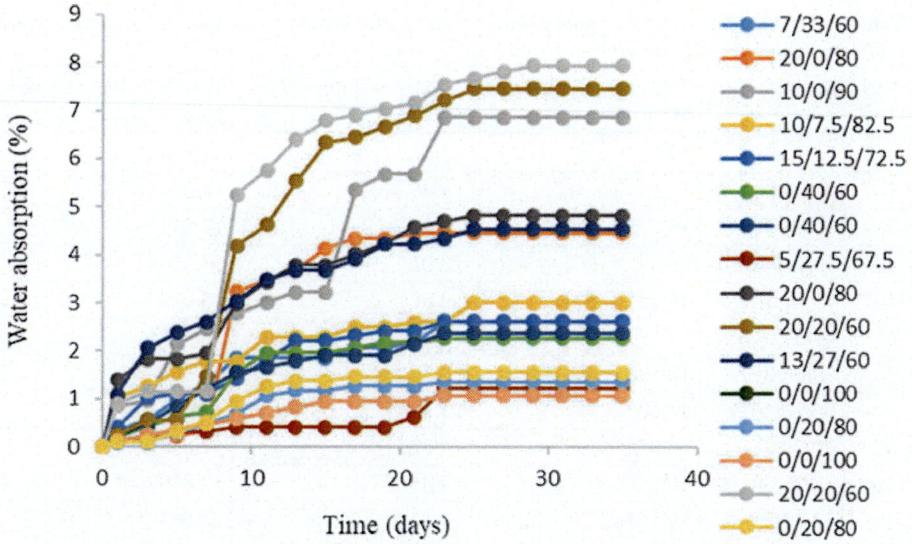

Fig. 3 Water absorption analysis for post-consumer glass and sawdust reinforced polyester composite

control (0/0/100 wt.%) having the least water absorption and the hybrid composition of 20/20/60 wt.% having the highest water absorption. The high water absorption for hybrid sample could be due to geometry parameters, reinforcement volume and type variation, and interface interactions with the matrix system [18]. Water absorption result ranged within 1.06 to 7.95% indicating a better result to Dotun et al. [19] absorption range of 6.955% to 13.37. Samples 10/0/90 and 20/0/80 reinforced with sawdust only had an increase in water absorption as the sawdust content increased due to the hydrophilic nature of natural fibres agreeing with Islam and Islam [20] that an increase in fibre loading increases the capacity of the composite to absorb water. The increasing water absorption due to sawdust filler loading proves the submission that natural fibres suffer absorption challenges [21]. Studies have reported that water absorption usually increases with immersion time although the rate of absorption decreases with increase in time [12]. The post-consumer glass and sawdust reinforced composites with low percentage water absorption of within 1.09–2.62% all had a saturation point at 23 days as against Alaneme et al. [12] water absorption study of polyester matrix composites reinforced with oil palm ash and oil palm fibre which attains equilibrium after 240 h (10 days) at which stage the composites attained saturation point.

Table 1 Tensile and water absorption property test result for post-consumer glass and sawdust reinforced polyester composites

S/N	Samples (%)	Tensile 1 (Dry sample) MPa	Water absorption (%)	Tensile 2 (Wet sample) MPa
1	7/33/60	19.762	2.62	22.836
2	20/0/80	19.538	4.46	18.061
3	10/0/90	19.275	6.86	14.738
4	10/7.5/82.5	24.699	3.01	14.016
5	15/12.5/72.5	15.227	2.62	15.015
6	0/40/60	14.783	2.27	10.414
7	0/40/60	10.198	2.38	9.36
8	5/27.5/67.5	11.603	1.23	12.633
9	20/0/80	16.072	4.82	17.859
10	20/20/60	10.248	7.46	12.846
11	13/27/60	19.161	4.55	22.845
12	0/0/100	28.868	1.06	29.339
13	0/20/80	12.154	1.36	12.861
14	0/0/100	28.868	1.06	29.339
15	20/20/60	11.884	7.95	11.14
16	0/20/80	9.416	1.56	15.147

Tensile Strength

Tensile property test result before undergoing weathering test ranged from 9.416 to 28.868 MPa with a range of 9.36 to 29.229 after undergoing degradation due to weathering effect (Table 1). Samples 7/33/60, 5/27.5/67.5, 20/0/80, 20/20/60, 13/27/60, 0/0/100, and 0/20/80 wt.% all had improved tensile strength after undergoing weathering test indicating high weathering resistance. While samples 10/0/20, 10/7.5/72.5, 15/12.5/72.5, and 0/40/60 wt.% reported lower tensile strength after water absorption indicating low resistance to weathering effect. Glass reinforced composites with some hybrid samples showed inferior tensile properties as compared to the control sample of 0/0/100% polyester matrix and sawdust reinforced composites. There was a decrease in tensile properties as the filler loading increase. This could be attributed to poor interfacial interactions of the reinforcement materials and matrix, as well as the fragile nature of glass material.

Weathering Resistance

Tensile strength and water absorption test results for sawdust, post-consumer glass, and hybrid reinforced polyester composites are presented in Table 1. Composite

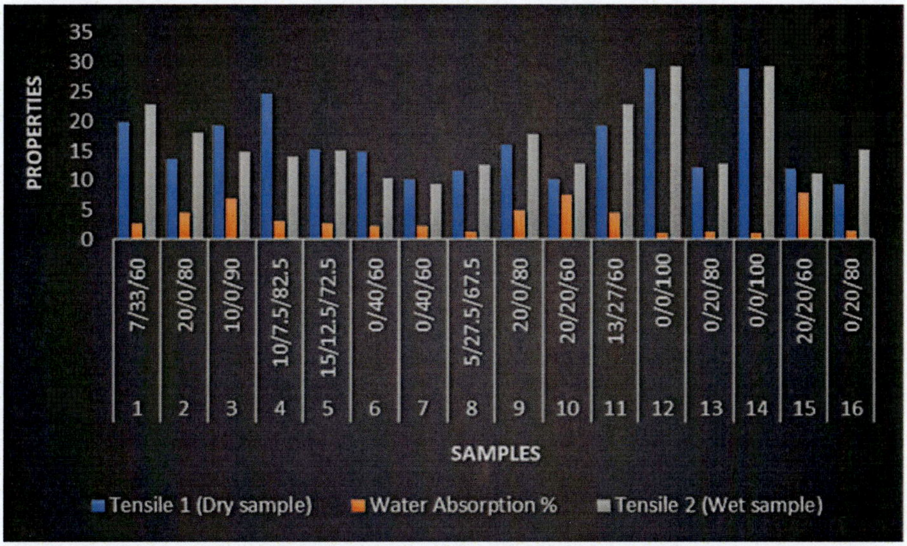

Fig. 4 Weathering resistance analysis for post-consumer glass and sawdust reinforced polyester composites

samples 5/27.5/67.5, 0/40/60, and 0/20/80 wt.% of hybrid and glass composition had the least WA of 1.23%, 1.91%, and 1.36% respectively due the good absorption properties of glass fibre. While sawdust reinforced sample 10/0/90 wt.% and hybrid sample 20/20/60 wt.% had the highest WA of 6.86 and 7.46% respectively. Samples 20/0/60 and 10/0/90 wt.% sawdust reinforced composites show low weathering resistance due to degradation after 35 days with a decrease in original tensile strength of 19.538 and 19.275 to 18.061 and 14.738 MPa respectively. This can be linked to high water absorption of 4.46, and 6.86% only next to samples 20/20/60 wt.% with the highest absorption of 7.46% and 7.95%. The increasing water absorption versus low weathering resistance proves that sawdust which is a natural fibre suffer more absorption challenges [21]. Hybrid samples 5/27.5/67.5, 7/33/60, and 20/20/60 wt.% with original tensile strength of 11.60, 19.76, and 10.25 MPa respectively had an improved tensile strength of 12.63, 22.84, 12.85 MPa respectively, indicating good resistance to weathering (see Fig. 4). This suggests that the composite material has the ability to withstand outdoor service conditions, agreeing with Krishnaiah [22] and Almansour [15] on the hybridization of natural and synthetic fibres for optimum performance properties.

Conclusion

Weathering resistance was studied by comparing changes in tensile strength before and after water absorption to determine the ability of post-consumer glass and sawdust reinforced polyester composites to withstand outdoor service conditions. The study revealed that samples 5/27.5/67.5, 0/40/60, and 0/20/80 wt.% of sawdust/glass/polyester hybrid compositions had the least WA of 1.23%, 1.91%, and 1.36% respectively while hybrid sample 20/20/60 had the highest WA of 7.46%. Hybrid samples 5/27.5/67.5, 7/33/60, and 20/20/60 wt.% with original tensile strength of 11.60, 19.76, and 10.25 MPa respectively had an improved tensile strength of 12.63, 22.84, and 12.85 MPa respectively. In general, samples 7/33/60, 5/27.5/67.5, 20/0/80, 20/20/60, 13/27/60, 0/0/100, and 0/20/80 wt.% all had improved tensile strength after undergoing weathering indicating high weathering resistance compared to samples 10/0/20, 10/7.5/72.5, 15/12.5/72.5, and 0/40/60 wt.% which reported lower tensile strength after water absorption indicating low resistance to weathering effect. Water absorption and tensile properties evaluated confirm that most of the hybrid and single reinforced composites reported good resistance to weathering. This suggests that the composite material has the ability to withstand outdoor service conditions, therefore, can be employed as particle board for outdoor application.

References

1. Abdullah AL-K, Abedin MZ, Beg MDH, Pickering KL, Khan MA (2006) Study on the mechanical properties of jute/glass fiber-reinforced unsaturated polyester hybrid composites: effect of surface modification by ultraviolet radiation. J Reinf Plast Compos 25(6):575–588
2. Egbo MK (2020) A fundamental review on composite materials and some of their applications in biomedical engineering. J King Saud Univ – Eng Sci. www.sciencedirect.com
3. Dawoud MM, Saleh HM (2018) Introductory chapter: background on composite materials. https://doi.org/10.5772/intechopen.80960
4. Sathishkumar TP, Naveen JA, Satheeshkumar S (2014) Hybrid fiber reinforced polymer composites – a review. J Reinf Plast Compos 33:454–471. https://doi.org/10.1177/0731684413516393
5. Madu OG, Nwankwojike BN, Ani OI (2018) Optimal design for rice husk-saw dust reinforced polyester ceiling board. Am J Eng Res (AJER). 7(6):11–16. www.ajer.org
6. Squire GN, Abhyankar A, Watson KJ (2008) Composite material manufactured from a binder system including waste powder coating powder. International application published under the patent cooperation treaty (Pct). International Publication Number PCT WO 2008/020768 Al
7. Hassanin AH, Hamouda T, Candan Z, Kilic A, Akbulutd T (2016) Developing high-performance hybrid green composites. Compos B Eng 92:384–394. https://doi.org/10.1016/j.compositesb.2016.02.051
8. Teja MS, Ramana MV, Sriramulu D, Rao CJ (2016) Experimental Investigation of Mechanical and Thermal properties of sisal fibre reinforced composite and effect of sic filler material. IOP Conf Ser: Mater Sci Eng 149:012095. https://doi.org/10.1088/1757-899X/149/1/012095
9. Corrosionpedia (2018) Weathering test - what does weathering test mean. https://www.corrosionpedia.com/definition/1396/weathering-test
10. Compound world (2011) Testing weathering resistance - Advances in testing weathering resistance. https://www.intertek.com/articles/advances-in-testing-weathering-resistance/
11. Markarian J (2011) Advances in testing weathering resistance. In: JE Casia composite show & conferences, Singapore. www.compoundingworld.com.
12. Alaneme KK, Oke SR, Omotoyinbo JA (2013) Water absorption characteristics of polyester matrix composites reinforced with oil palm ash and oil palm fibre. Usak Univ J Mater. http://uujms.usak.edu.tr
13. Hassan BB, Awopetu OO (2019) Production and characterization of particle boards from common agro wastes in Nigeria. Int J Innov Sci Res Technol. www.ijisrt.com

14. Pickering KL, Aruan E, Fendy MG, Le TM (2015) A review of recent developments in natural fibre composites and their mechanical performance. https://doi.org/10.1016/j.compositesa
15. Almansour H (2018) Mechanics and design of structures using advance composites. Mechanics 12, Retrieved Sept. 3 from https://www.researchgate.net/profile/Gaurav-5/publication/331985123
16. Singh H, Singh, T (2019) Effect of fillers of various sizes on mechanical characterization of natural fiber polymer hybrid composites: a review. Mater Today: Proc. 18. https://doi.org/10.1016/j.matpr.2019.07.560
17. Kim H (2014) Hybrid composites with natural fibres. A Master dissertation. School of metallurgy and materials, college of engineering and physical sciences, University of Birmingham
18. Motoc DL (2015) Tailoring the effective properties of hybrid polymer based composite materials. Habilitation Thesis, Transilvania University of Brașov, pp 13–25
19. Dotun AO, Adediran AA, Oluwatimilehin AC (2018) Physical and mechanical properties evaluation of particle board produced from saw dust and plastic waste. Int J Eng Res Afr. https://doi.org/10.4028/www.scientific.net/JERA.40.1
20. Islam MN, Islam MS (2015) Characterization of chemically modified sawdust-reinforced recycled polyethylene composites. J Thermoplast Compos Mater 28(8):1135–1153. https://doi.org/10.1177/0892705713503671
21. Rodriguez-Castellanos W, Rodrigue D (2016) Production and characterization of hybrid polymer composites based on natural fibers. Department of chemical engineering and CERMA, Université Laval, Quebec, QC, Canada, IntechOpen. https://doi.org/10.5772/64995
22. Krishnaiah P (2017) Development of polylactide and polypropylene composites reinforced with sisal fibres and halloysite nanotubes for automotive and structural engineering applications. PhD thesis, University of Nottingham. http://eprints.nottingham.ac.uk/43498/1/PhD%20Thesis_Prakash%20Krishnaiah.pdf

Preparation of FeMnAlSiC Powder by CO_2-Steel Slag Cooperative Electro Deoxidation

Zhenwei Jing, Xiaofei Xing, Ju Meng, Hongyan Yan, Hui Li, and Jinglong Liang

Abstract The resource utilization of CO_2 and steel slag is an urgent problem that needs to be solved. In order to solve the problem of high added value utilization of CO_2-steel slag, thermodynamic calculations were carried out for the preparation of FeMnAlSiC by molten salt electrolysis. Results showed that during 800–1000 °C, the voltage range of −2.51 to −2.39 V could ensure the reduction of slag, and the reduction order was Fe, Mn, Si, and Al. CO_2 was continuously introduced into NaCl–$CaCl_2$, existed in the form of CO_3^{2-}. C will appear under the voltage of −2.77 to −2.51 V, and then reacted with Fe, Mn, Al, Si in the cathode to synthesize FeMnAlSiC powder. In conclusion, considering energy consumption, thermodynamic conditions for the synthesis of metal elements in the cathode and ensuring that molten salt was not electrolyzed, the electrolysis temperature and voltage were 900 °C and −3 V respectively. Under this condition, FeMnAlSiC powder could be prepared theoretically, and the order of alloying in theory was Mn–Si, Fe–Si, Si–C, Mn–C, Al–Si–C, Fe–C.

Keywords CO_2-Steel slag · Resource utilization · Electroreduction · FeMnAlSiC

Introduction

CO_2 emission of China's steel industry is second only to the electric power industry by released data from Carbon Emission Accounts & Database (CEADs); it is about 18% of the total CO_2 emission [1]. At present, there are three main coping strategies for a large amount of CO_2 [2]. The first is to develop new technologies and new energy. The second is to develop CO_2 storage technology. The third is to recycle CO_2 as a valuable resource. Developed countries such as Norway, Sweden, and Canada

Z. Jing · J. Meng · H. Yan (✉) · H. Yan (✉) · H. Li · J. Liang
College of Metallurgy and Energy, North China University of Science and Technology, Tangshan 063210, China
e-mail: yanhy@ncst.edu.cn

X. Xing
Tangshan Research Institute, Beijing Institute of Technology, Tangshan 063004, China

B. Li et al. (eds.), *Advances in Powder and Ceramic Materials Science 2023*, The Minerals, Metals & Materials Series, https://doi.org/10.1007/978-3-031-22622-9_11

have built CCS projects for CO_2 and have mature CO_2 storage technologies. Carbon Cycle International (CRI) Iceland has developed technology to convert CO_2 and H_2 into renewable methanol. Liquid CO_2 and dry ice plants have been established in California in the United States and Japan to produce dry ice and liquefied CO_2 [3]. In China, high-value chemicals and fuels are produced from CO_2 in the coal chemical industry. In the field of calcium-containing and alkali-containing sewage treatment, adding CO_2 to sewage can achieve the purpose of removing calcium ions, softening water quality, and neutralizing. The reaction achieves the purpose of treating alkali-containing sewage. In the iron and steel smelting industry, carbon dioxide is added to the sewage to form insoluble calcium carbonate, thus remove calcium ions and reduce the total hardness of the wastewater. In addition, building materials products are prepared by carbonating steel slag with CO_2. It can be seen from the above that the current research on the utilization of CO_2 as a resource is mainly based on the bulk utilization of low added value, and there are few studies on the direction of high added value.

In addition, with the increase of China's steel production, the solid wastes generated in the process of steel making are also increasing. The existing steel slag in China has exceeded 200 million tons [4], but its comprehensive utilization rate is only about 22%. Most of the steel slag is still stored in the open air [5]. This situation not only occupies a large amount of land, but also pollutes the surrounding environment by the chemical composition in the steel slag. At the beginning of the twentieth century, some developed countries have begun to study the comprehensive utilization technology of steel slag, and the recycled steel slag was applied to road engineering, agricultural fertilizer, building materials, and civil engineering. However, how to use steel slag with high added value and environmental protection is still an urgent problem to be solved in the global metallurgical industry.

At present, the utilization of CO_2 and steel slag is mainly carried out in a single aspect. Although CO_2 has been used for carbonation treatment of steel slag, it is also used as low added value products such as building materials [6–8]. Existing CO_2 electroreduction studies have shown that CO_2 can be electrolyzed into C and O_2 under the action of electric field [9–11], and the research on the preparation of alloy powder by electroreduction of blast furnace slag and steel dust by molten salt electrolysis had also achieved certain results [12–14]. Therefore, the green and energy-saving molten salt electrolysis method can be considered to reduce the CO_2 and steel slag. The C produced by the CO_2 reduction reacts with the high-value Fe, Mn, Al, Si obtained by the electric reduction of the steel slag to form FeMnAlSiC powder. This scheme will provide a new idea and theoretical basis for realizing the resource utilization of the steel slag. In the future, CO_2-steel slag collaborative resource utilization will bring great social and economic effects. Therefore, thermodynamic analysis of CO_2-steel slag synergistic electroreduction was carried out to prepare high value-added FeMnAlSiC powder, theoretically analyzing the feasibility of CO_2-steel slag synergistic electroreduction treatment. Analyzing the interaction between CO_2 and various components in steel slag to determine the sequence of reduction of Fe, Mn, Al, Si, and C, obtained the reaction temperature and potential of electroreduction preparation.

Reaction Thermodynamics

In this experiment, CO_2 and steel slag cooperative electroreduction was adopted. C produced by CO_2 reduction reacts with Fe, Mn, Al, Si obtained by the reduction of steel slag to form FeMnAlSiC powder and achieve the cooperative utilization of steel slag and CO_2. Therefore, thermodynamic calculations are needed to verify the feasibility of the scheme and determine the experimental conditions.

Molten Salt System

Molten salt plays a decisive role in the whole process of electro deoxidation. As the medium of ion transfer, molten salt determines the speed of electrolysis and the energy consumption of electrolysis process. NaCl–$CaCl_2$ molten salt system was selected in this experiment because it is relatively cheap, non-toxic and harmless; the temperature required for melting is relatively low. These characteristics make it suitable for industrial production and application. Figure 1 shows the binary phase diagram of NaCl–$CaCl_2$. When the molar ratio of NaCl and $CaCl_2$ is 0.48:0.52, the melting point of the system is the lowest, which is 504 °C. So the electrolysis temperature should be at least greater than 504 °C. During the electrolysis process, CO_2 is continuously introduced into the molten salt and exists in the form of CO_3^{2-}. CO_3^{2-} will continuously combine with free Ca^{2+} in molten salt to form $CaCO_3$, and exist in this form as the C source in the preparation of FeMnAlSiC powder. Figure 2 is the theoretical decomposition voltage diagram of NaCl and $CaCl_2$. This molten salt system has a wide potential window, and the decomposition voltage of NaCl is lower when the temperature is above 800 °C. In order to protect molten salt from electrolysis, the voltage applied in this system shall not exceed −3.24 to −3.11 V within the temperature range of 800–1000 °C.

Thermodynamic Calculation of Electrochemical Reaction

The elements such as Fe, Mn, Al, and Si in the steel slag can be considered to exist in the form of oxides such as FeO, MnO_2, Al_2O_3, and SiO_2 in the electrolytic system. In order to prepare FeMnAlSiC powder, it is necessary to control the voltage to electrolyze the metal oxides and $CaCO_3$. After calculating the standard Gibbs free energy of the electrochemical reactions involved in the process by using FactSage 7.3, the standard theoretical electrolytic voltage $E^{\ominus}$ of each compound was calculated using formula (1).

$$\Delta G^{\ominus} = -n\mathrm{F}E^{\ominus} \tag{1}$$

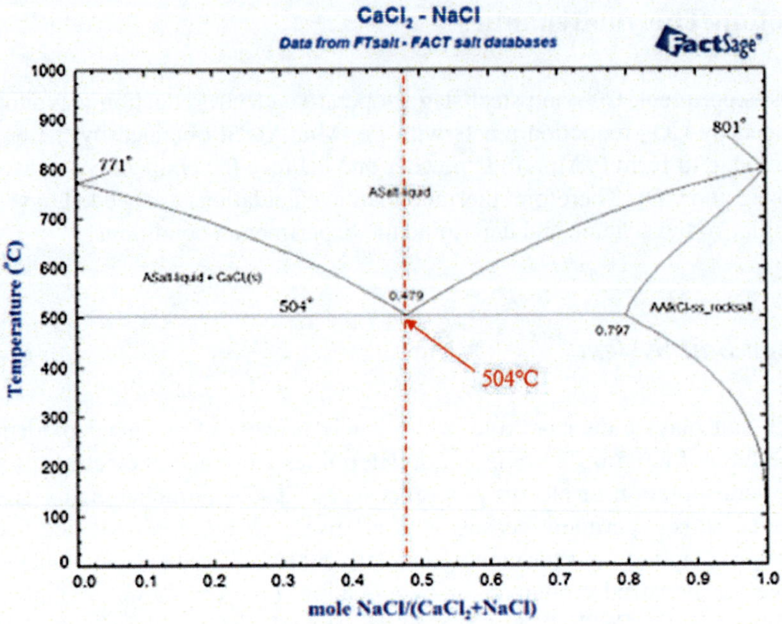

Fig. 1 NaCl–CaCl$_2$ molten salt phase diagram

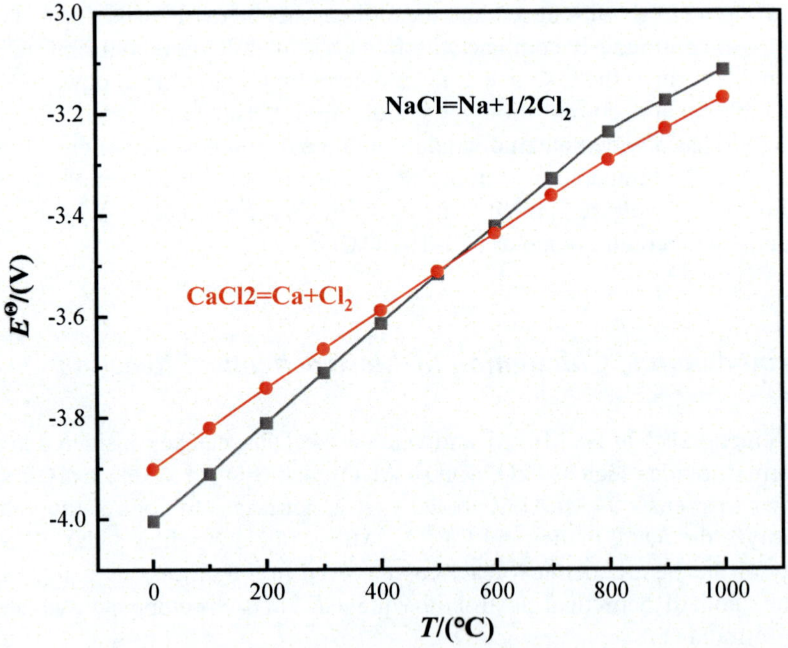

Fig. 2 Decomposition voltage of NaCl and CaCl$_2$ at 0–1000 °C

where $\Delta G^{\ominus}$ was the standard Gibbs free energy (kJ mol^{-1}), $E^{\ominus}$ was the theoretical decomposition voltage in the standard state (V), F was the Faraday constant (96,485 C mol^{-1}), and n was the number of electrons gained or lost in the reaction equation.

The theoretical decomposition voltage of each compound at 0–1200 °C was calculated to obtain Fig. 3.

Figure 3 shows that within the temperature range of 500–1000 °C, the order of metal oxide reduction is FeO, MnO, SiO_2, and Al_2O_3. The required theoretical decomposition voltage is less than 3 V, which is within the potential window of NaCl–$CaCl_2$ molten salt system. The theoretical voltage of $CaCO_3$ electrolysis to C and CaO is between FeO and MnO. When various oxides of the cathode are deoxidized, a large amount of O^{2-} will be transferred from the cathode surface to the molten salt. Due to the limitation of the ion transfer rate of the molten salt, some O^{2-} may not leave the cathode in time, the high oxygen activity is likely to cause O^{2-} combine with Ca^{2+} to generate CaO. The generated CaO will react with Al_2O_3 in the cathode to generate $Ca_{12}Al_{14}O_{33}$, as shown in formula (2). In order to ensure the purity of the product, it is also necessary to control the voltage to remove it.

$$12CaO + 7Al_2O_3 = Ca_{12}Al_{14}O_{33} \quad (2)$$

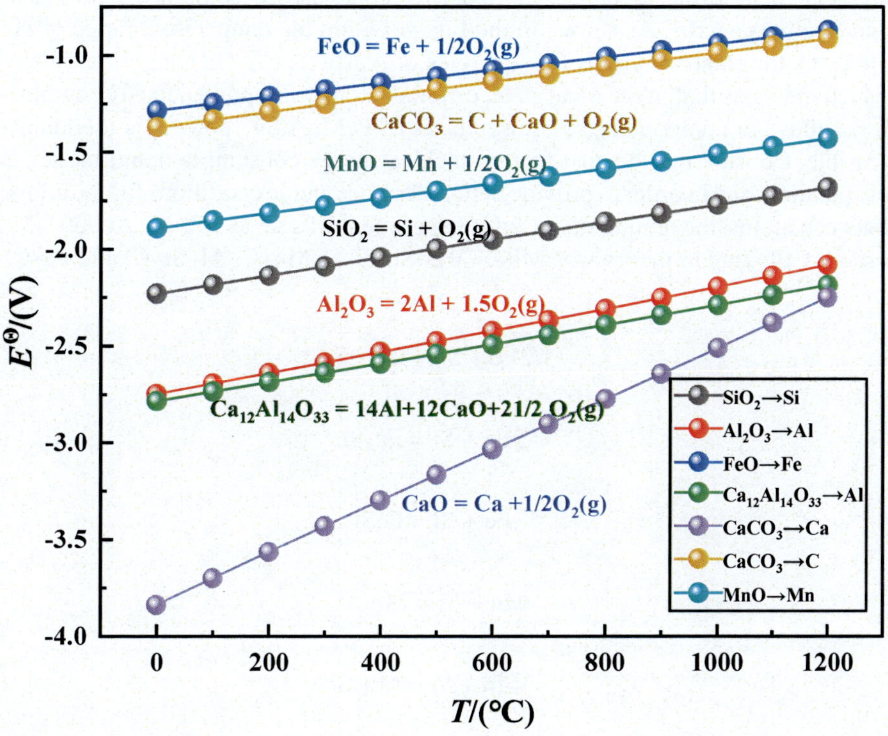

Fig. 3 Calculated results between $E^{\ominus}$ and T at 0–1200 °C

Since CO_2 is continuously supplied, $CaCO_3$ will be continuously electrolyzed into CaO. If the voltage does not reach the electrolysis voltage of CaO, the amount of CaO in the molten salt may increase continuously, causing Ca^{2+} in the molten salt be consumed. Moreover, the solubility of NaCl–$CaCl_2$ system to CaO is limited. With the progress of electrolysis, it may lead to the deposition of CaO, which will adversely affect the preparation process. According to the above calculation results, the voltage in this experiment should be controlled within the range of greater than the CaO electrolysis voltage and less than the molten salt electrolysis voltage. In order to speed up the process of electro deoxidation, the experimental voltage can be appropriately increased. The voltage was finally determined to be −3 V.

Alloying Reaction

From the theoretical voltage calculation in the previous section, it can be concluded that the order of reduction is FeO, $CaCO_3$, MnO, SiO_2, and Al_2O_3. The binary system alloys may form after electrolysis include: Fe–Si, Fe–Al, Mn–Si, and Al–Si. Among them, Fe, Mn, and Si are reduced earlier than Al. Therefore, the reduced Al can be rapidly combined with the adjacent Fe and Si atoms or the formed Fe–Si and Mn–Si to further alloying. In theory, C atoms can be combined with Fe, Mn, and Si atoms nearby. As shown in the Fig. 4, within the temperature range of 800–1000 °C, the standard Gibbs free energies of reactions (3)–(12) are all less than 0, which indicates that in the ideal state, different atoms can spontaneously combine to form alloys or compounds, so the synthesis of FeMnAlSiC powder is theoretically feasible. Considering the cost factors such as energy consumption and molten salt evaporation, and in order to provide better thermodynamic conditions for the alloying between atoms, the temperature of this experiment was set at 900 °C. At 900 °C, the order of alloying in theory was Mn–Si, Fe–Si, Si–C, Mn–C, Al–Si–C, and Fe–C.

$$1/2Fe + Si = 1/2FeSi_2 \tag{3}$$

$$3Fe + Si = Fe_3Si \tag{4}$$

$$Fe + Si = FeSi \tag{5}$$

$$Mn + Si = Mn_3Si \tag{6}$$

$$3Mn + Si = Mn_3Si \tag{7}$$

$$3Fe + C = Fe_3C \tag{8}$$

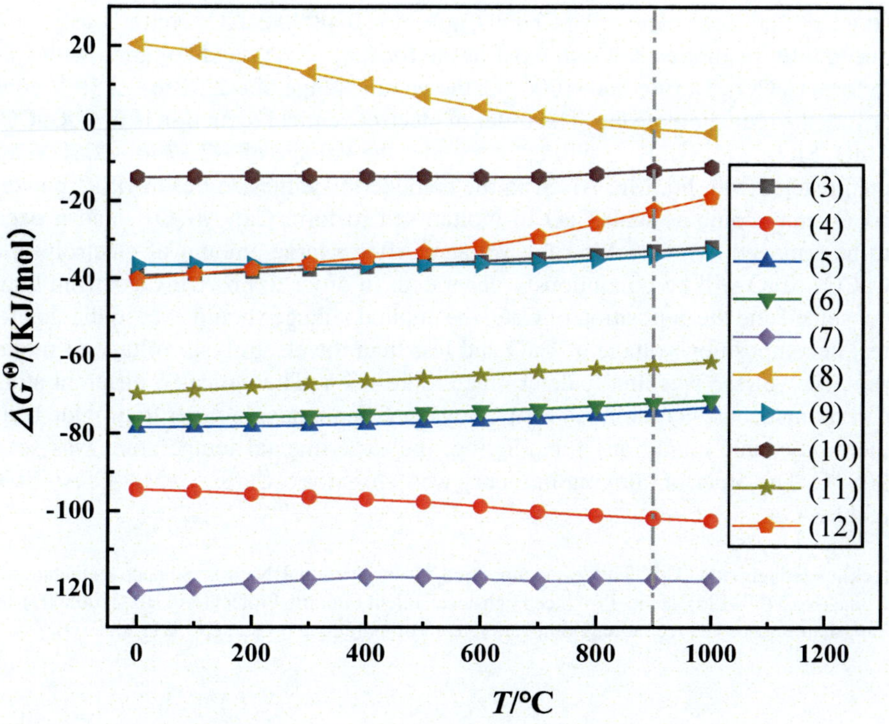

Fig. 4 Calculated results between ΔG^{Θ} and T at 0–1000 °C

$$7/3Mn + C = 1/3Mn_7C_3 \tag{9}$$

$$3Mn + C = Mn_3C \tag{10}$$

$$Si + C = SiC \tag{11}$$

$$C + 4/3Al + 1/3SiC = 1/3Al_4SiC_4 \tag{12}$$

Conclusion

The thermodynamic analysis of CO_2-steel slag synergistic electroreduction was carried out to prepare FeMnAlSiC powder. The electroreduction conditions were obtained, and the CO_2-steel slag synergistic electroreduction treatment was feasible. NaCl–$CaCl_2$ molten salt system with low price and wide potential window was adopted. The molar ratio of NaCl and $CaCl_2$ was 0.48:0.52. CO_2 continuously introduced into the molten salt will exist in the form of ${CO_3}^{2-}$ and combine with Ca^{2+} to form $CaCO_3$. During 800–1000 °C, the voltage range of −2.51 to −2.39 V could ensure the reduction of slag. The order of electrochemical reduction is FeO, $CaCO_3$, MnO, SiO_2, and Al_2O_3. C will appear under the voltage of −2.77 to −2.51 V, and then reacted with Fe, Mn, Al, Si in the cathode to synthesize FeMnAlSiC powder. Al_2O_3 may combine with CaO in molten salt to form $Ca_{12}Al_{14}O_{33}$, and it needs to be removed by controlling the voltage. After a large amount of electrolysis of $CaCO_3$, CaO will be continuously generated. In order to prevent Ca^{2+} from being consumed and the deposition of CaO, the applied voltage should exceed the theoretical decomposition voltage of CaO and less than the electrolysis voltage of molten salt. The voltage was finally determined to be −3 V. Theoretically, different atoms can spontaneously combine to form alloys or compounds. In order to provide better thermodynamic conditions for alloying, the experimental temperature was set at 900 °C. The order of alloying in theory was Mn–Si, Fe–Si, Si–C, Mn–C, Al–Si–C, and Fe–C.

Acknowledgements This work was supported by the National Natural Science Foundation of China (No.51804126), Hebei Province High-level Talent Funding Project (No. B2022005007), and Tangshan science and technology innovation team training plan project (No. 21130207D).

References

1. Shan Y, Huang Q, Guan D et al (2020) China CO_2 emission accounts 2016–2017. Sci Data 7(1):54
2. Hongyan Y, Chao L, Xiaojun H et al (2018) Utilization status of CO_2 in iron and steel industries. Nonferr Metal Sci Eng 9(06):26–30
3. Chauvy R, Weireld GD (2020) CO_2 utilization technologies in Europe: a short review. Energy Technol 8(12):1–17
4. Cailiang P, Xueqing Y, Jieguang S et al (2020) Research status and development trend of comprehensive utilization of steel slag. Brick Tile 3:77–80
5. Liu Z, Zong Y, Feng H et al (2015) Influence of ferrum on crystallization and microstructure of steel slag based glass-ceramics. Trans Indian Ceram Soc 74(1):29–34
6. Chun W, Jingling Y, Guilin Z et al (2012) Development and present situation of high value utilization technology of steel slag. Iron Steel Scrap China 1:42–53
7. Wang R, Jin P, Dong H et al (2020) Effect of moist content on the bio-carbonated steel slag bricks. Constr Build Mater 269:121313
8. Song Q, Guo MZ, Wang L et al (2021) Use of steel slag as sustainable construction materials: a review of accelerated carbonation treatment. Resour Conserv Recycl 173:105740
9. Peterson AA, Norskov JK (2012) Activity descriptors for CO_2 electroreduction to methane on transition-metal catalysts. J Phys Chem Lett 3(2):251–258
10. Tan X, Yu C, Ren Y et al (2021) Recent advances in innovative strategies for the CO_2 electroreduction reaction. Energy Environ Sci 14(2):765–780
11. Zhang X, Sun X, Guo SX et al (2019) Formation of lattice-dislocated bismuth nanowires on copper foam for enhanced electrocatalytic CO_2 reduction at low overpotential. Energy Environ Sci 12(4):1334–1340

12. Jihong D, Li Q, Shenghong Y et al (2009) Preparation of TiV Alloy by Electrolysis-Deoxidization in Molten Salt. Rare Met Mater Eng 100(12):1739–1746
13. Baichuan H, Ziyue L, Dongfang J et al (2020) Comprehensive utilization of blast furnace slag containing titanium. Multipurp Utili Miner Resour 6:1–6
14. Shi J, Qiu Y, Yu B et al (2022) Titanium extraction from titania-bearing blast furnace slag: a review. JOM 74(2):654–667

Thermodynamic Analysis of BN Prepared by Electrodeposition BN Power

Ju Meng, Chao Luo, Zhenwei Jing, Hongyan Yan, Hui Li, and Jinglong Liang

Abstract BN has a wide application in the fields of anticorrosion, adsorption, and lubrication, due to its excellent physical and chemical properties. With the development of electrochemical methods, BN was prepared by electrodeposition of B_2O_3 in NaCl–$CaCl_2$ and N_2 atmosphere, which provided a new idea for the preparation of BN. The results of thermodynamic analysis showed that the theoretical decomposition voltage of B_2O_3 was −1.73 to −1.66 V at 800–1000 °C. The electrodeposition process was accompanied by chemical synthesis of borate with B_2O_3, O^{2-}, and metal ions, the formation sequence was $Na_2B_4O_7$, $Na_2B_2O_4$, $Ca_3B_2O_6$, $Ca_2B_2O_5$, CaB_2O_4, and decomposition voltage of these borates was −2.19 to −1.75 V. Spontaneous reaction between N_2 and B generated BN. Considering factors such as energy consumption, electrolysis conditions of raw materials and borate, and molten salt state, it was determined that the electrolysis temperature and electrolysis voltage were 800–1000 °C and −2.19 to −1.75 V, respectively. BN powder can be prepared theoretically.

Keywords BN power · Thermodynamic analysis · Electrodeposition · NaCl–$CaCl_2$ molten salt

Introduction

Boron nitride (BN) is a crystal composed by nitrogen and boron atoms. Because structure is similar to graphite, BN is called white graphite [1]. BN is an inorganic non-metallic material with excellent performance, such as good mechanics, thermal conductivity, wear resistance, barrier properties, hydrophobicity, and adsorption. BN is widely used in aerospace [2], high-temperature lubricants [3], sensors [4],

J. Meng · Z. Jing · H. Yan (✉) · H. Li · J. Liang
College of Metallurgy and Energy, North China University of Science and Technology, Tangshan 063210, China
e-mail: yanhy@ncst.edu.cn

C. Luo
Hesteel Group Tangsteel Company, Tangshan 063000, China

B. Li et al. (eds.), *Advances in Powder and Ceramic Materials Science 2023*, The Minerals, Metals & Materials Series, https://doi.org/10.1007/978-3-031-22622-9_12

adsorption [5], and many other fields. At present, the main methods for preparing BN are high-temperature and high-pressure synthesis method [6], chemical vapor deposition method [7], hydrothermal synthesis method [8], precursor method [9], and molten salt method [10]. Zhao [11] found that the agglomeration of borax at high temperature caused the uneven mixing of raw material, affected the degree of heat transfer, and caused defects such as large fluctuations in product structure. Stehle [12] studied the effect of temperature and heat treatment of raw materials on the preparation of BN by chemical vapor deposition. The results showed that the reaction rate of BN formation was slow, the whole preparation process needed to be precisely controlled, and industrialization was relatively difficult. Above methods have some disadvantages, such as low reaction rate, high raw material price, and high energy consumption. The molten salt method has certain advantages. The reaction speed is usually higher, and the preparation process is relatively simple. However, various pivotal influencing factors such as temperature and raw material ratio need to be determined by experiment. Therefore, it is necessary to explore a new way of preparing BN materials.

Electrochemical deposition method [13] is a method that the current passes through the migration of positive and negative ions in the electrolyte solution and deposits on the electrode under the action of an external electric field. It has been widely used in the preparation of new materials in recent years [14, 15]. Based on the principle of molten salt method and electrodeposition method, the electrodeposition method to prepare BN can be explored. It provides a new direction for BN preparation. Under a certain voltage condition, the boron-containing ions are electrodeposited on the stainless steel cathode to form B element, and further nitridation is performed to synthesize BN. Compared with the molten salt method, the electrodeposition method further increases the reaction rate under the action of the electric field, the reaction temperature decreases, and the operation process is simple. It is valuable to carry out this research work. In order to analyze the feasibility of preparing BN by electrodeposition, a theoretical analysis of thermodynamics is required. In this paper, using B_2O_3 as raw material and NaCl–$CaCl_2$ as molten salt, a series of reactions in the electrodeposition process and nitrification process were analyzed, the sequence of electrodeposition containing boron ions was determined, and the voltage and temperature conditions of BN electrodeposition were obtained.

Reaction Thermodynamics

Thermodynamic Calculation of Boron Oxide and Molten Salt

In order to obtain boron ions in the electrodeposition process, boron oxide was mixed with NaCl–$CaCl_2$ molten salt and dissolved at a rising temperature. During the dissolution process, boron oxide and molten salt undergo a series of chemical reactions.

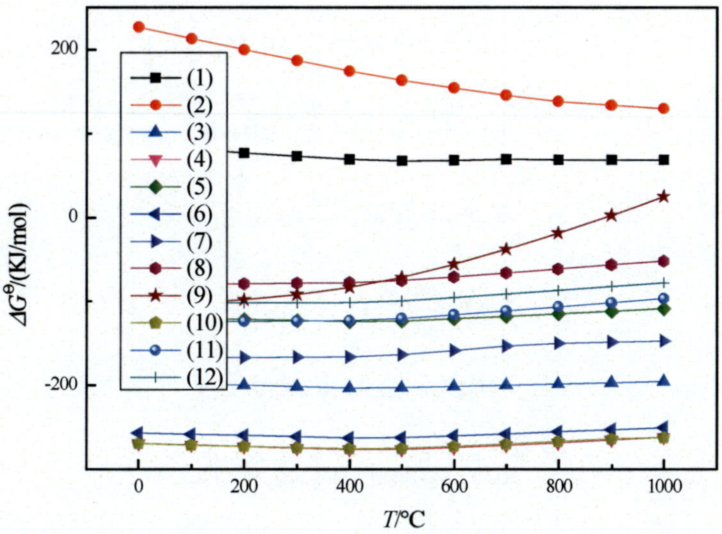

Fig. 1 Standard gibbs free energies for chemical reactions between boron oxide and molten salts and borates at different temperatures

Reaction (1), (2) was the chemical reaction of boron oxide and molten salt, and reaction (3)–(12) was the chemical reaction of borate formation. According to Fig. 1, the standard Gibbs free energy of the reaction between boron oxide and molten salt was positive. It indicated that no chemical reaction occurred between them at 0–1000 °C. In the melting process, boron oxide can spontaneously combine with free Na^+, Ca^{2+}, and O^{2-} ions in molten salt to form various borates. The standard Gibbs free energy was more negative, and the reaction occurred more likely. Therefore, the borate formation sequence was $NaBO_2$, $Ca_3B_2O_6$, $Ca_2B_2O_5$, $Na_2B_4O_7$, CaB_2O_4, $Na_2B_6O_{10}$, $Na_2B_8O_{13}$, and CaB_4O_7. $Ca_2B_6O_{11}$ cannot be generated above 800 °C. After the boron oxide was dissolved in the molten salt, a variety of borates were formed. The boron ions in the molten salt include B^{3+} and borate ions.

$$\frac{7}{2}B_2O_3 + 3NaCl = \frac{3}{2}Na_2B_4O_7 + BCl_3(g) \tag{1}$$

$$5B_2O_3 + 6CaCl_2 = 3Ca_2B_2O_5 + 4BCl_3(g) \tag{2}$$

$$B_2O_3 + 2CaO = Ca_2B_2O_5 \tag{3}$$

$$B_2O_3 + Na_2O = Na_2B_2O_4 \tag{4}$$

$$B_2O_3 + CaO = CaB_2O_4 \tag{5}$$

$$B_2O_3 + 3CaO = Ca_3B_2O_6 \quad (6)$$

$$B_2O_3+\frac{1}{2}Na_2O = \frac{1}{2}Na_2B_4O_7 \quad (7)$$

$$B_2O_3+\frac{1}{2}CaO = \frac{1}{2}CaB_4O_7 \quad (8)$$

$$B_2O_3+\frac{2}{3}CaO = \frac{1}{3}Ca_2B_6O_{11} \quad (9)$$

$$B_2O_3 + Na_2O = 2NaBO_2 \quad (10)$$

$$B_2O_3+\frac{1}{3}Na_2O = \frac{1}{3}Na_2B_6O_{10} \quad (11)$$

$$B_2O_3+\frac{1}{4}Na_2O = \frac{1}{4}Na_2B_8O_{13} \quad (12)$$

Electrochemical Reactions of Boron Containing Substances

The thermodynamic behavior of electrochemical reaction of boron in molten salt can be obtained by the theoretical decomposition voltage $E^{\ominus}$ of different substances. HSC6.0 software was used to calculate the standard Gibbs free energy of the reaction, and $E^{\ominus}$ was calculated by the following formula

$$\Delta G^{\ominus} = -n\mathrm{F}E^{\ominus}$$

where $\Delta G^{\ominus}$ was the standard Gibbs free energy (kJ mol^{-1}), $E^{\ominus}$ was the theoretical decomposition voltage in the standard state (V), F was the Faraday constant (96,485 C mol^{-1}), and n was the number of electrons gained or lost in the reaction equation.

In molten salt, boron oxide was B_2O_3, calcium borate, sodium borate and other boron containing substances after dissolution. A series of electrochemical reactions occurred when these boron containing substances were energized. The electrochemical reactions were mainly divided into the electrochemical reaction of boron oxide and the electrochemical reaction of borate.

(1) Electrochemical reaction of boron oxide
It can be seen from Fig. 2 that the theoretical decomposition voltage of reaction (13) was most positive. It indicated that the reaction of boron oxide electrolysis to elemental B was most likely to occur. The occurrence of reactions (14)–(19) consumed boron oxide, and the experimental conditions should be controlled

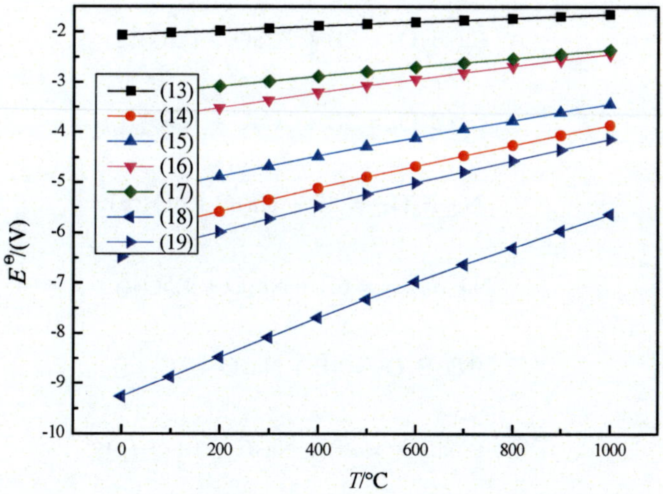

Fig. 2 Theoretical decomposition voltage of boron oxide at different temperatures

to avoid the occurrence of these side reactions. Therefore, the decomposition voltage of boron oxide at 800–1000 °C was −1.73 to −1.66 V.

$$B_2O_3 = 2B + 1.5O_2(g) \tag{13}$$

$$2B_2O_3 = 4BO(g) + O_2(g) \tag{14}$$

$$2B_2O_3 = B + 3BO_2(g) \tag{15}$$

$$2B_2O_3 = 2B_2O_2(g) + O_2(g) \tag{16}$$

$$B_2O_3 = B_2O(g) + O_2(g) \tag{17}$$

$$B_2O_3 = BO(g) + BO_2(g) \tag{18}$$

$$3B_2O_3 = B_2O(g) + 4BO_2(g) \tag{19}$$

(2) Electrochemical reaction of borate

$$CaB_4O_7 = 4B + CaO + 3O_2(g) \tag{20}$$

$$CaB_2O_4 = 2B + CaO + 1.5O_2(g) \tag{21}$$

$$Ca_2B_2O_5 = 2B + 2CaO + 1.5O_2(g) \tag{22}$$

$$Ca_3B_2O_6 = 2B + 3CaO + 1.5O_2(g) \tag{23}$$

$$Na_2B_8O_{13} = 8B + Na_2O + 6O_2(g) \tag{24}$$

$$Na_2B_6O_{10} = 6B + Na_2O + 4.5O_2(g) \tag{25}$$

$$Na_2B_4O_7 = 4B + Na_2O + 3O_2(g) \tag{26}$$

$$2NaBO_2 = 2B + Na_2O + 1.5O_2(g) \tag{27}$$

$$4\,NaBO_2 = 4BO(g) + 2\,Na_2O + O_2(g) \tag{28}$$

$$2\,NaBO_2 = B_2O(g) + Na_2O + O_2(g) \tag{29}$$

$$Na_2B_4O_7 = 4BO(g) + Na_2O + O_2(g) \tag{30}$$

$$Na_2B_4O_7 = 2\,B_2O(g) + Na_2O + 2O_2(g) \tag{31}$$

$$Na_2B_6O_{10} = 6BO(g) + Na_2O + 1.5O_2(g) \tag{32}$$

$$Na_2B_6O_{10} = 3\,B_2O(g) + Na_2O + 3O_2(g) \tag{33}$$

$$Na_2B_8O_{13} = 8BO(g) + Na_2O + 2O_2(g) \tag{34}$$

$$Na_2B_8O_{13} = 4\,B_2O(g) + Na_2O + 4O_2(g) \tag{35}$$

$$2Ca_3B_2O_6 = 4BO(g) + 6CaO + O_2(g) \tag{36}$$

$$Ca_3B_2O_6 = B_2O(g) + 3CaO + O_2(g) \tag{37}$$

$$2Ca_2B_2O_5 = 4BO(g) + 4CaO + O_2(g) \tag{38}$$

$$Ca_2B_2O_5 = B_2O(g) + 2CaO + O_2(g) \tag{39}$$

$$CaB_2O_4 = B_2O(g) + CaO + O_2(g) \quad (40)$$

$$2CaB_2O_4 = 4BO(g) + 2CaO + O_2(g) \quad (41)$$

$$CaB_4O_7 = 4BO(g) + CaO + O_2(g) \quad (42)$$

$$CaB_4O_7 = 2\ B_2O(g) + CaO + 2O_2(g) \quad (43)$$

The decomposition voltage distribution was more regularity; borate decomposition reaction was more likely to react. When the decomposition products were BO and other products, the decomposition voltage was very negative, negative to the decomposition voltage of molten salt. When the decomposition products were B_2O gas and other products, the decomposition voltage was close to the molten salt decomposition voltage, and the decomposition voltage range was −3.22 to −2.50 V at 800–1000 °C. When the decomposition products were B and other products, the decomposition voltage was relatively positive and the reaction is the first. According to Fig. 3, the borate electrolysis sequence was CaB_4O_7, $Na_2B_8O_{13}$, $Na_2B_6O_{10}$, CaB_2O_4, $Na_2B_4O_7$, $Ca_2B_2O_5$, $Ca_3B_2O_6$, and $NaBO_2$. In order to avoid the occurrence of side reactions, the decomposition voltage range of borate at 800–1000 °C was −2.19 to −1.75 V.

(3) Electrochemical reactions of molten salts
The selection of molten salt system has an important influence on the current efficiency of electrolytic boron oxide. The common molten salt system is NaCl–KCl, NaCl and $CaCl_2$, etc. In particular, NaCl–$CaCl_2$ system has the advantages of low price, large specific heat capacity, low viscosity, wide operating temperature range, good stability, and high decomposition voltage. Table 1 showed that when the molar ratio of $CaCl_2$–NaCl was 52.9:47.1, the eutectic point of the molten salt system was 494 °C. It was lower than that of the single electrolyte system and had the advantages of low energy consumption and higher safety [16]. The electrolysis temperature was higher, the kinetic conditions were better, but the high temperature leaded to anodization and volatilization of molten salt. Therefore, the electrolytic temperature should be 150–200 °C higher than the eutectic point of molten salt system. In this experiment, the electrolytic temperature range was 800–1000 °C, and the molar ratio of $CaCl_2$–NaCl was 52.9:47.1.

$$2NaCl = 2Na + Cl_2(g) \quad (44)$$

$$CaCl_2 = Ca + Cl_2(g) \quad (45)$$

According to Fig. 4, the theoretical decomposition voltages of NaCl and $CaCl_2$ were −3.24 V to −3.17 V and −3.48 V to −3.41 V at 800 °C–1000 °C, respectively.

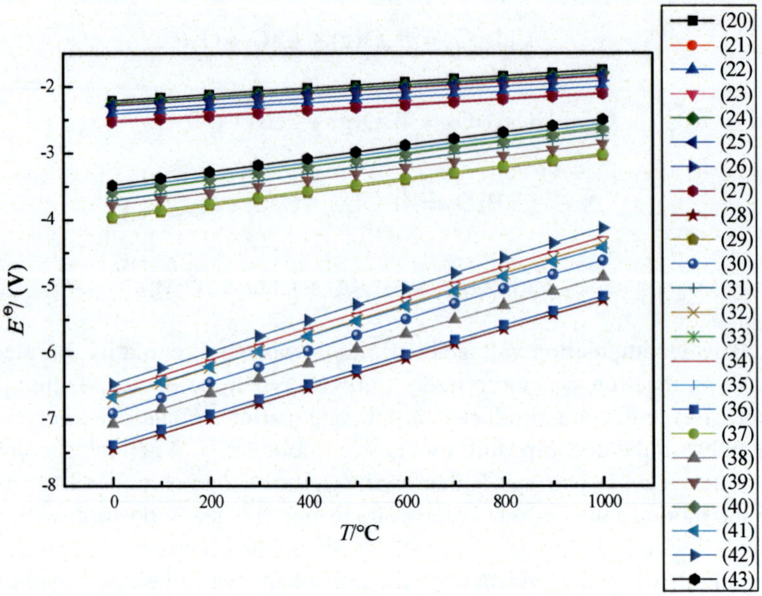

Fig. 3 Theoretical decomposition voltage of borate at 0–1000 °C

Table 1 Common chloride system ratio and co-melting point

Molten salt system	Composition (%)	Melting point (°C)
NaCl–iCl	27.0–73.0	553
NaCl–KCl	50.0–50.0	657
NaCl–$CaCl_2$	47.1–52.9	494
NaCl–$BaCl_2$	48.5–51.5	648
KCl–LiCl	58.0–42.0	348
KCl–$CaCl_2$	75.0–25.0	615
KCl–$BaCl_2$	55.0–45.0	655

The added voltage should not exceed −3.17 V, otherwise the molten salt was decomposed. It can be seen that the potential window of NaCl–$CaCl_2$ system was wide, and the decomposition voltages of NaCl and $CaCl_2$ were significantly larger than the theoretical decomposition voltages of raw materials and products. The melting point of boron oxide was close to 450 °C, which was conducive to the fusion of boron oxide and molten salt system, and promoted the full electrodeposition. Combining the electrolytic voltage of molten salt, boron oxide, and borate, the theoretical decomposition voltage of 800–1000 °C was −2.19 to −1.75 V.

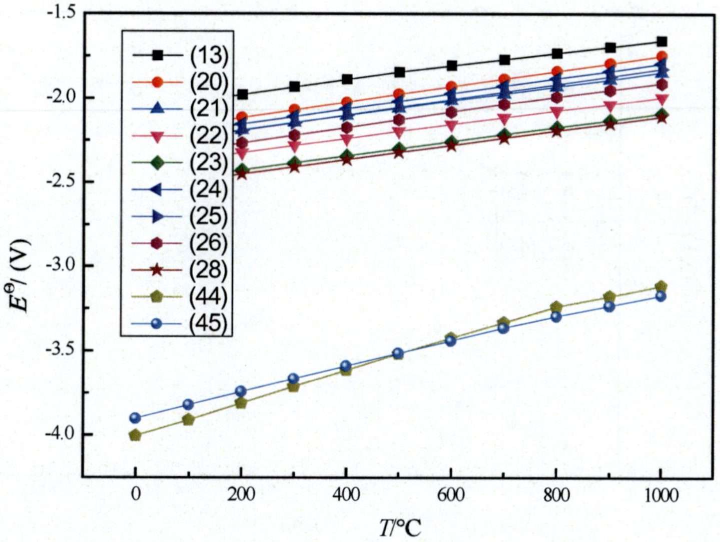

Fig. 4 Decomposition voltage of boron source and molten salt at 0–1000 °C

Alloys and Nitriding Reactions

After the electrodeposition of B on the stainless steel cathode, the alloy reaction occurred between B and stainless steel at the phase interface at the experimental temperature, forming Fe-B alloy. In addition, after the end of the electrodeposition reaction, nitrogen was injected into the reaction system to nitride B to form BN. The specific alloy reaction and nitriding reaction were shown in Eqs. (46)–(48).

$$Fe + B = FeB \tag{46}$$

$$2Fe + B = Fe_2B \tag{47}$$

$$2B + N_2(g) = 2BN \tag{48}$$

According to Fig. 5, the electrodeposited B spontaneously reacted with Fe to form FeB and Fe_2B on the interface between B and stainless steel cathode. Fe_2B was preferentially generated by FeB, and neither FeB nor Fe_2B reacted with molten salt. The nitriding reaction of B can proceed spontaneously at room temperature, and the actual kinetic process needed a higher reaction temperature to improve the reaction rate. Gao Xiaoju [17] showed that the B nitriding reaction rate was suitable at 800–1000 °C. At the same time, according to thermodynamic calculation, other substances in the system cannot react with N_2 under the condition of power failure.

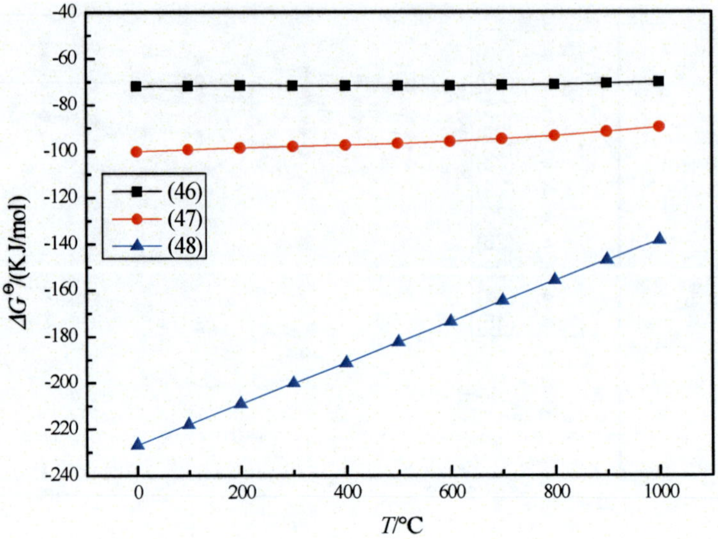

Fig. 5 Standard Gibbs free energy of Fe-B alloy reaction and B-nitride reaction at different temperatures

Conclusions

BN prepared by electrodeposition BN power with raw materials of B_2O_3 was theoretically feasible. In the process of B_2O_3 melting, borate was used as boron source to participate in the electrodeposition of B. Boron oxide dissolved in molten salt to form some borates. Combined with thermodynamic calculation of boron source, the electrolytic order was B_2O_3, CaB_4O_7, $Na_2B_8O_{13}$, $Na_2B_6O_{10}$, CaB_2O_4, $Na_2B_4O_7$, $Ca_2B_2O_5$, $Ca_3B_2O_6$, and $NaBO_2$. NaCl–$CaCl_2$ molten salt system was selected. The ratio of NaCl and $CaCl_2$ was 47.1:52.9, and the electrolytic temperature was 800–1000 °C. The decomposition voltage range of borate was −2.19 to −1.75 V. FeB and Fe_2B were formed on the interface between B and stainless steel cathode. The nitriding reaction temperature was 800–1000 °C.

Acknowledgements This work was supported by the National Natural Science Foundation of China (No.51804126), Hebei Province High-level Talent Funding Project (No. B2022005007) and Tangshan science and technology innovation team training plan project (No. 21130207D).

References

1. Zhi C, Bando Y, Terao T et al (2010) Towards thermoconductive, electrically insulating polymeric composites with boron nitride nanotubes as fillers. Adv Func Mater 19(12):1857–1862
2. Xu Z, Chen Y, Li W et al (2018) Preparation of boron nitride nanosheet-coated carbon fibres and their enhanced antioxidant and microwave-absorbing properties. RSC Adv 8
3. Corthay S, Firestein KL, Kvashnin DG et al (2021) Elevated-temperature high-strength h-BN-doped Al2014 and Al7075 composites: experimental and theoretical insights. Mater Sci Eng A 809(8):140969
4. Wang W, Deng P, Liu X et al (2020) A $CsPbBr_3$ quantum dots/ultra-thin BN fluorescence sensor for stability and highly sensitive detection of tetracycline. Microchem J 162(22):105876
5. Huda M N, Kleinman L (2006) h-BN monolayer adsorption on the Ni (111) surface: a density functional study. Phys Rev B Condens Matter 74(7)
6. Zhi C, Bando Y, Tan C et al (2005) Effective precursor for high yield synthesis of pure BN nanotubes. Solid State Commun 135(1–2):67–70
7. Yuan SD, Wang JF, Zhang GX et al (2009) Investigation on effect of the CVD temperature onto synthesis of boron nitride nanosphere. J Wuhan Univ Technol
8. Wei F, Hao XP, Yang YL et al (2006) Effects of nitrogen source on the boron nitride nanocrystals prepared by solvothermal synthesis route. J Synth Cryst
9. Zhang X, Chang Q, Wei X et al (2019) Preparation and characterization of high crystallinity hexagonal boron nitride. Diam Abras Eng
10. Ye LF, Zhao L, Liang F et al (2015) Facile synthesis of hexagonal boron nitride nanoplates via molten-salt-mediated magnesiothermic, reduction. Ceram Int 10:14941–14948
11. Zhao HM, Xu XW, Li YP et al (2002) Reaction mechanism of $Na_2B_4O_7$ with NH_4Cl in hBN synthesis. Chin J High Press Phys 16(4):4
12. Stehle Y, Meyer HM, Unocic RR et al (2015) Synthesis of hexagonal boron nitride monolayer: control of nucleation and crystal morphology. Chem Mater 27(23):8041–8047
13. Liu W, Kamiko M, Yamada I et al (2022) Electrochemical deposition of amorphous cobalt oxides for oxygen evolution catalysis. RSC Adv 12(14):8731–8736
14. Sun Z et al (2021) Electrochemical deposition of Cu metal–organic framework films for the dual analysis of pathogens. Anal Chem 93(25):8994–9001
15. Manoharan S et al (2021) Electrochemical deposition of vertically aligned tellurium nanorods on flexible carbon cloth for wearable supercapacitors. Chem Eng J (2):129548
16. Jia L, Yan H Y, Li H et al (2020) Preparation and refining of metal hafnium. Multipurp Utiliz Mineral Resour (1):6
17. Gao XJ, Wang HJ, Zhang DH (2009) Preparation of h-BN ceram ICS by reaction sintering. Aerosp Mater Technol 39(1):4

Use of Ceramic Waste in Different Percentages as a Replacement of the Fine Aggregate in Mortars

M. G. P. Cherene, G. C. Xavier, A. R. G. Azevedo, and S. N. Monteiro

Abstract With the growth of the population and, consequently, of civil construction, there was an increase in the consumption of natural resources, generating a large amount of solid waste and contributing to various environmental impacts. The application of solid waste in civil construction aims to reduce environmental pollution and the construction of sanitary landfills. Thus, this work presents the use of ceramic waste as a substitute for fine aggregate in proportions of 10, 20, and 30% in mortars. To evaluate its use, tests of consistency index, density in the fresh state, tensile strength in flexion and axial compression, tensile bond strength, and microstructural analysis were performed using the Scanning Electron Microscopy (SEM) test. Through the results, it is concluded that the best proportion used is 10% of ceramic waste, which indicated results superior to the reference mix, presenting smaller pores, resulting in high strengths. At 28 days, it obtained 2.91 MPa of tensile strength and 9.40 MPa of compressive strength.

Keywords Ceramic waste · Multipurpose mortars · Environmental impacts

M. G. P. Cherene
UENF-Laboratory of Advanced Materials, State University of Northern Rio de Janeiro, Rio de Janeiro 28013-602, Brazil
e-mail: marianagpc1@yahoo.com.br

G. C. Xavier · A. R. G. Azevedo (✉)
Civil Engineering Laboratory, UENF-State University of Northern Rio de Janeiro, Rio de Janeiro 28013-602, Brazil
e-mail: afonso@uenf.br

G. C. Xavier
e-mail: gxavier@uenf.br

S. N. Monteiro
IME-Military Engineering Institute, Department of Material Sciences, Rio de Janeiro 22290-270, Brazil

B. Li et al. (eds.), *Advances in Powder and Ceramic Materials Science 2023*, The Minerals, Metals & Materials Series, https://doi.org/10.1007/978-3-031-22622-9_13

Introduction

Currently, the world generates a large amount of waste, due to rapid population growth, industrial and agricultural activities that pollute and degrade our environment, through the release of pollutants, and construction of landfills for waste disposal [1]. The drastic increase in the amount of industrial agricultural waste has caused serious environmental problems in our societies [1, 2].

In this way, issues related to environmental preservation, waste reduction, and recycling have occupied a prominent place in the discussions held by society to achieve a sustainable development model for the planet. In Brazil, the concern with the reuse of solid waste is relatively recent, but the recycling of construction materials is at an advanced stage of discussion by various sectors of civil society in order to seek solutions to minimize the problem [3, 4].

According to Dimitriou et al. [5], waste consumptions are in accordance with essential environmental requirements for approaches, waste prevention, reuse of waste materials, reduce landfill area, reuse energy from waste, and save natural resources. While alternative materials such as waste are used in many fields of application, there is a need to address their technical characteristics, financial aspects, and environmental implications.

In Campos dos Goytacazes, in the north of the state of Rio de Janeiro, Brazil, the red ceramic manufacturing process, which involves manufacturing, transport, and storage, generates approximately 19 thousand tons of waste per month [6]. It is estimated that about 30% of the daily production of ceramic industries occurs in the form of waste of defective parts. Among the main defects observed in the blocks, which generate the loss in question, the cracks due to shrinkage can be mentioned, in general derived from the proportionally inappropriateness of the clays and the high moisture content in the manufacture [5, 6].

In this way, the use of ceramic waste can contribute to the management of solid waste in red ceramics, reducing costs with raw material extraction, including increasing the useful life of the deposits, adding economic value to the waste, and correct destination of the waste [7].

In this context, it should be noted that the main objective of this work is to evaluate the application of ceramic waste as a partial substitute for sand, evaluating through workability, density in the fresh state, flexural tensile and compressive strength, tensile adherence strength, and X-ray diffraction (XRD). The main gap in this research is the use of mortar without the use of lime, which greatly reduces the cost, unlike other authors, who use lime-based mortar.

Materials and Methods

The ceramic waste was collected from a ceramic industry located in the municipality of Campos dos Goytacazes. This residue was obtained from the disposal of ceramic blocks that went through the firing process at temperatures between 600 and 630 °C [8, 9]. Soon after, the ceramic waste was properly ground in a ball mill. Cement and sand taken from Paraíba do Rio Sul were also used. The mix used was 1:6 (Portland cement: natural sand) by mass, one of the most commonly used mixed mortar ratios for mortar production [10]. The quantities established in Table 1 were used for the preparation of the mortars according to the recommendations of procedures of the Brazilian standard [11]. The tests carried out were the consistency index, which consists of determining the amount of water necessary to maintain the spread between 260 mm ± 5 mm and mass density in the fresh state, following the recommendations of procedures of the Brazilian standard [12], showing that all mixtures had density higher than the reference trace (CPII00). The evaluated traits and consistency indices are shown in Table 1.

For the tests in the hardened state, prismatic specimens with dimensions 40 × 40 × 160 mm were molded, and each test in the hardened state had three specimens according of procedures of the Brazilian standard [13]. The curing procedure of the specimens was carried out for 28 days at room temperature. The flexural tensile and compressive strength tests were carried out in according of procedures of the Brazilian standard [13] and were carried out with the aid of an INSTRON 5582 press with a maximum capacity of 10 tons. The load used in the test was 50 ± 10 N/s at a speed of 1 mm/min in the tensile strength test, and in the compressive strength test, a load of 500 ± 50 N/s at a speed of 10 mm was used /min. The tensile bond strength test was also carried out in according to procedures of the Brazilian standard [14], using a masonry wall, after which the distribution of 12 specimens was carried out in each mortar trace. The distribution was made so that the specimens were spaced from each other, in addition to the corners and corners, by at least 50 mm. To obtain the specimens, cuts were first made with a hole saw, 50 mm in diameter, to a depth of 1 mm, without reaching the substrates (masonry wall). Finally, microstructural characterization tests were carried out at 28 days, through X-ray diffraction (XRD) performed in a diffractometer with an angular step of 0.02° with an interval of 1 s and an angle of 2θ ranging from 5° to 70°.

Table 1 Proportions, workability, and density results

Mixtures	Cement (g)	Sand (g)	Ceramic waste (g)	Consistency index (mm)	W/A	Density (g/cm^3)
CPII00	286	1710	–	258	1,36	1,90
CPII10	286	1540	171	265	1,47	1,99
CPII20	286	1360	342	260	1,67	1,97
CPII30	286	1200	513	255	1,96	1,91

Results

Figure 1 shows the results of flexural tensile strength at 28 days of curing and shows that the mixture with the highest strength is CPII10 with 3.24 MPa and that all mixtures had higher strength than the reference mixture (CPII00). In general, the behavior of the loss of flexural tensile strength related to the increase of the ceramic waste can be caused by the physical characteristics of the ceramic waste, such as lower densities than sand and higher absorptions [15]. The study by Gayarre et al. [10] found a range of 1.5–2.5 MPa of strength, and as the amount of residue increased, there was a gain in strength. Mohit and Sharifi [16] partially replaced the cement with ceramic waste and as the amount of ceramic waste increased, the strength decreased, however, the strength of the 10% replacement was higher than the reference mixture, the strength remained at the same level range from 2.86 to 3.84 MPa, a result that corroborates the findings of this research.

Figure 2 shows the results of compressive strength at 28 days of curing and indicates that the blend with the highest compressive strength is CPII10 with 9.40 MPa. In general, all traces had higher strength than the reference mixture (CPII00); this fact also occurred in the flexural tensile strength test. In general, the behavior of the loss of compressive strength related to the increase of the ceramic waste can be caused by the physical characteristics of the ceramic waste, similar to the causes of the tensile strength in bending, such as lower densities than sand and higher absorptions [15].

Gayarre et al. [10] found, through compressive strength, the range from 5.5 to 8.5 MPa, and it was possible to notice that not always when the amount of residue had been added, the strength increased. According to Jain et al. [17], the use of ceramic waste above the 20% level led to the dilution of the C–S–H forming compounds and,

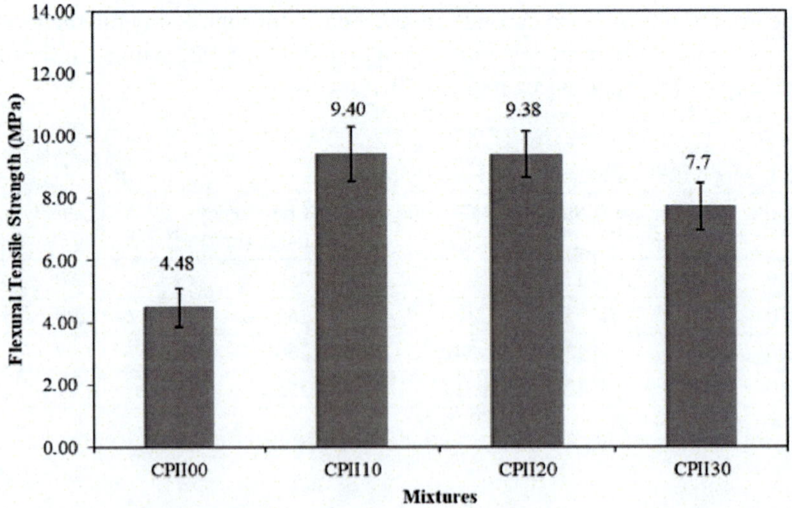

Fig. 1 Flexural tensile strength results

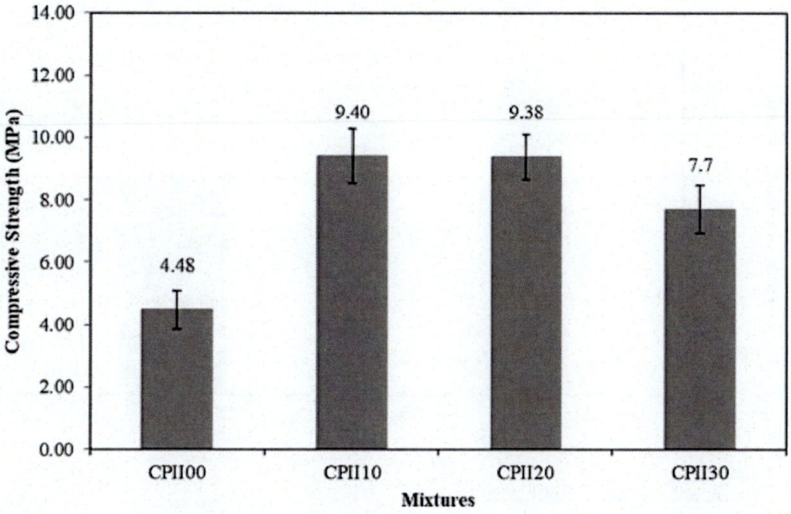

Fig. 2 Compressive mechanical strength results

consequently, to the delay in setting and hardening of the cementitious composites. This was also accompanied by a drop in compressive strength.

Figure 3 shows how the mortars behaved after laying on the masonry at 28 days, and it is possible to notice that the mixtures CPII00 and CPII10 showed no change, while the other mixtures showed the fissure pathology; this fact can be explained by the results found in the consistency index test, in which the mixtures with higher percentages of ceramic waste have a greater amount of water, directly interfering with the workability and adhesion of the mortar. The crack pathology appears while the mortar was not yet completely dry, that is, the mixtures that obtained cracks had rapid evaporation of water before the end of setting, which may configure plastic shrinkage or have lost water to the substrate. Therefore, the CPII20 and CPII30 mixtures are not recommended for use. This property is directly related to the flexural tensile and compressive strength.

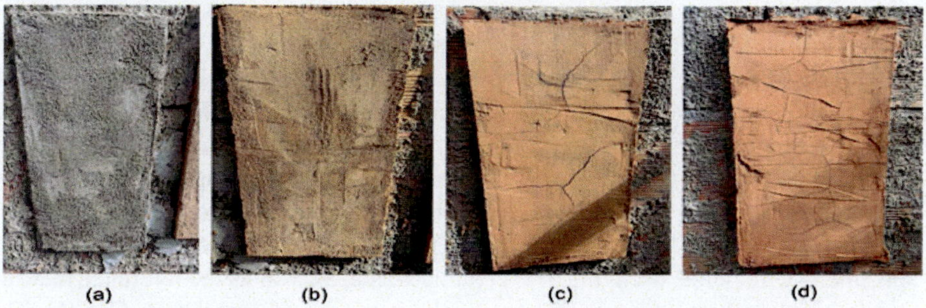

Fig. 3 Appearance of mortars in masonry: **a** CPII00; **b** CPII10; **c** CPII20; **d** CPII30

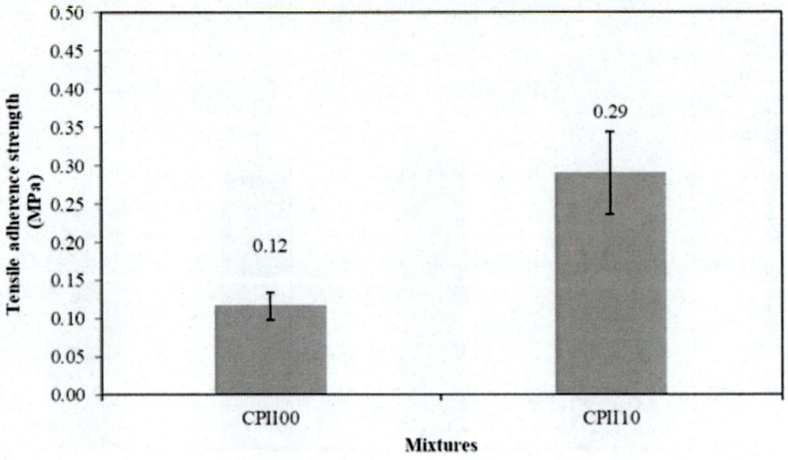

Fig. 4 Tensile adherence strength results

Figure 4 shows the tensile strength results. It is observed that the ceramic waste promoted an increase in the tensile adherence strength. The CPII10 mixture obtained the highest strength with 0.29 MPa, superior strength to the reference mixture (CPII00). Based on the minimum values of according to the Brazilian standard [18], the mortar with 10% of the ceramic waste (CPII10) can be used as an internal coating, as long as the strength is greater than 0.20 MPa. However, it cannot be used as an external coating. The increase in resistance that occurred between the mortar and the ceramic substrate, which increased as a function of the use of the ceramic waste, shown in Fig. 4, can be explained by the reduction in porosity. Thus, the increase in adhesion is not explained simply by the use of ceramic waste, but together with the other results indicated in this work as resistances [19].

Figure 5 shows the results of Scanning Electron Microscopy (SEM); it was possible to notice that the traces have the presence of cement hydration reactions, as they have C–S–H (hydrated calcium silicate), ettringite, and portlandite. Regarding the reference mixture (CPII00), it can be seen that these mixtures have a microstructure with large pores and the hydration products are not very adherent to the sand grains. The mixture with the use of 10% of the ceramic waste has smoother surfaces, assuming their filler effect. The ceramic waste has a high amount of aluminum oxide, which can contribute to the formation of C–S–H. In the study by Mohammadhosseini et al. [4] in the analysis of the SEM in their mortar identified that the C–S–H crystals formed in the matrix with 40% of ceramic waste made the mortar denser due to the pozzolanic reaction between silica and $Ca(OH)_2$ during cement hydration, resulting in to increase the strength of the mortar. Results are consistent with those found in this research. Mohit and Sharifi [16] also performed the SEM analysis of the specimens with better strength, 10% ceramic waste, as well as the reference mixture and found that the two mixtures present the presence of amorphous C–S–H. The mortar with the residue has smaller pores than the reference mortar, that is, the reference mortar had less hydration and was not enough to fill the capillary pores of the

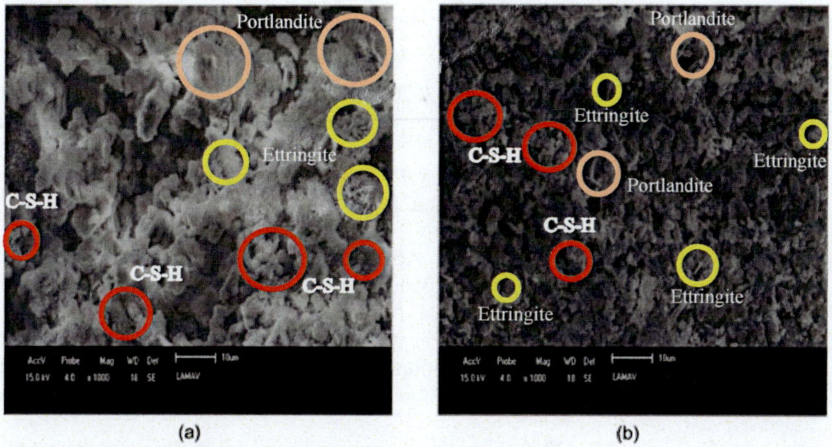

Fig. 5 Scanning electron microscopy (SEM) results

mortar. Therefore, the ceramic waste helped to improve hydration. Pitarch et al. [20] found the presence of portlandite, ettringite, and C–S–H in the reference mixture and the mixture containing 25% ceramic residue at 28 days, the residue particles were completely surrounded by reaction products, and hydrated calcium silicate was the main product of the reaction observed; the same occurred in Fig. 5.

Conclusion

- The best flexural tensile and compressive strength was obtained with mortars containing 10% of the ceramic waste. All mixtures containing ceramic waste are recommended, as they had higher strengths than the reference mixtures.
- Regarding the tensile adherence strength, it was observed that the mixtures CPII20 and CPII30 showed cracks. The characteristics that presented this pathology were the characteristics with the highest w/c ratio. The trait with the best performance was CPII10, the same occurred for flexural and tensile strength. Through the Scanning Electron Microscopy (SEM) test, these results were confirmed, which indicated that the reference mortar is more porous than the traces that have ceramic waste.
- Through Scanning Electron Microscopy (SEM), it was identified that the mortars have portlandite, C–S–H, and ettringite, and the mortars with the ceramic waste have fewer pores than the reference mortars.
- Therefore, it is concluded that it is possible to use the ceramic waste in mortars for laying blocks and covering ceilings and walls. In particular, mortar containing 10% of the ceramic waste is recommended, since these were the traits that presented the best results in the properties.

References

1. Ngayakamo BH, Bello A, Onwualu AP (2020) Development of eco-friendly fired clay bricks incorporated with granite and eggshell wastes. Environ Challeng 1:100006. https://doi.org/10.1016/j.envc.2020.100006
2. Xin Y, Mohajerani A, Kurmus H, Smith JV (2021) Possible recycling of waste glass in sustainable fired clay bricks: a review. Int J Geo 78:57–64. https://doi.org/10.21660/2021.78.Gx260
3. Silva VM, Góis LC, Duarte JB, Silva JB, Acchara W (2014) Incorporation of ceramic waste into binary and ternary soil-cement formulations for the production of solid bricks. Mater Res 17:326–331. https://doi.org/10.1590/S1516-14392014005000014
4. Mohammadhosseini H, Lim NHAS, Tahir MM, Alyousef R, Alabduljabbar H, Samadi M (2019) Enhanced performance of green mortar comprising high volume of ceramic waste in aggressive environments. Constr Build Mater 212:607–617. https://doi.org/10.1016/j.conbuildmat.2019.04.024
5. Dimitriou G, SavaP PMF (2018) Enhancing mechanical and durability properties of recycled aggregate concrete. Constr Build Mater 158:228–235. https://doi.org/10.1016/j.conbuildmat.2017.09.137
6. Azevedo ARG, França BR, Alexandre J, Marvila MT, Zanelato EB, Xavier GC (2018) Influence of sintering temperature of a ceramic substrate in mortar adhesion for civil construction. J Build Eng 19. https://doi.org/10.1016/j.jobe.2018.05.026
7. Zhao Y, Gao J, Liu C, Chen X, Xu Z (2020) The particle-size effect of waste clay brick powder on its pozzolanic activity and properties of blended cement. J Clean Prod 242:118521. https://doi.org/10.1016/j.jclepro.2019.118521
8. Monteiro SN, Vieira CMF, Ribeiro MM, Silva FAN (2007) Red ceramic industrial products incorporated with oily wastes. Constr Build Mater 21:2007–2011. https://doi.org/10.1016/J.CONBUILDMAT.2006.05.035
9. Barreto ES, Stafanato KV, Marvila MT, Azevedo ARG, Ali M, Pereira RML, Monteiro SN (2021) Clay ceramic waste as Pozzolan constituent in cement for structural concrete. Materials 14:2917. https://doi.org/10.3390/ma14112917
10. Gayarre FL, Boadella ÍL, Pérez CL, López MS, Cabo DA (2017) Influence of the ceramic recycled agreggates in the masonry mortars properties. Constr Build Mater 457–461. https://doi.org/10.1016/J.CONBUILDMAT.2016.12.021
11. ABNT NBR 16541:2016 (2016) Mortar for laying and covering walls and ceilings—Preparation of the mixture for testing; Brazilian Association of Technical Standards; ABNT: São Paulo, Brazil. (In Portuguese)
12. ABNT NBR 13278:2005 (2005) Mortar for laying and covering walls and ceilings—Determination of mass density and content of incorporated air; Brazilian Association of Technical Standards; ABNT: São Paulo, Brazil. (In Portuguese)
13. ABNT NBR 13279:2005 (2005) Mortar for laying and covering walls and ceilings—Determination of tensile strength in bending and compression; Brazilian Association of Technical Standards; ABNT: São Paulo, Brazil. (In Portuguese)
14. ABNT NBR 13258:2021 (2021) Mortar for coating walls and ceilings—Determination of potential tensile bond strength; Brazilian Association of Technical Standards; ABNT: São Paulo, Brazil. (In Portuguese)
15. Cabrera-Covarrubias FG, Gómez-Soberón JM, Almaral-Sánchez JL, Arredondo-Rea SP, Corral-Higuera R (2015) Mechanical properties of mortars containing recycled ceramic as a fine aggregate replacement. Revista de la Construcción 14:22–29. https://doi.org/10.4067/S0718-915X2015000300003
16. Mohit M, Sharif Y (2019) Thermal and microstructure properties of cement mortar containing ceramic waste powder as alternative cementitious materials. Constr Build Mater 223:643–656. https://doi.org/10.1016/j.conbuildmat.2019.07.029

17. Jain P, Gupta R, Chaudhary S (2022) Literature review on the effect of using ceramic waste as supplementary cementitious material in cement composites on workability and compressive strength. Materials. https://doi.org/10.1016/j.matpr.2022.03.453
18. ABNT NBR 13281:2005 (2005) Mortar for laying and covering walls and ceilings—Requirements; ABNT: São Paulo, Brazil. (In Portuguese)
19. Azevedo ARG, Alexandre J, Zanelato EB, Marvila MT (2017) Influence of incorporation of glass waste on the rheological properties of adhesive mortar. Constr Build Mater 59–368. https://doi.org/10.1016/J.CONBUILDMAT.2017.04.208
20. Pitarch AM, Reig L, Tomás AE, Forcada G, Soriano L, Borrachero MV, Payá J, Monzó JM (2021) Pozzolanic activity of tiles, bricks and ceramic sanitary-ware in eco-friendly Portland blended cements. J Clean Prod 279:123713. https://doi.org/10.1016/j.jclepro.2020.123713

Part II
Powder Materials Processing and Fundamental Understanding

Combustion Synthesis of ZrC-TiC Composite Nanoparticle by Self-Propagating High Temperature Synthesis (SHS) in ZrO_2–TiO_2–Mg/Al–C System

Mehmet Bugdayci and Ozan Coban

Abstract In this study, ZrC–TiC composite nanoparticle was synthesized by SHS method using oxide raw materials, carbon black, and Mg and Al reductants. For SHS processes, composite charge stoichiometries were optimized for Mg usage, and the usage of Al and Mg was compared. The stoichiometries of the chemicals used in the processes applied to remove undesired by-products and the most accurate process steps were determined for the purification of the SHS product. A novel route was established for purification of SHS product obtained by Al usage as reductant. Characterization was performed with XRD analysis. The results showed that commercial purity ZrC–TiC powder with high surface area could be synthesized by using both reductants. The results revealed that Mg is a better reductant, but Al with lower cost when compared to Mg is also a suitable reductant, although it increases the process steps.

Keywords Nanoparticle synthesis · Ceramic composite · Advanced ceramics · Zirconium carbide · Titanium carbide

Introduction

Zirconium carbide has superior properties such as very high melting temperature, irradiation resistance, hardness, and wear resistance. However, sintering difficulty, low fracture toughness, and flexure strength limit its usage [1–5]. The studies carried out have revealed that the production of composite with TiC reduces the sintering temperature and also increases the fracture toughness [6]. Therefore, the synthesis

M. Bugdayci
Chemical Engineering Department, Yalova University, 77200 Yalova, Turkey

O. Coban (✉)
Metallurgical and Materials Engineering Department, Istanbul Technical University, Istanbul, Turkey
e-mail: ozan.coban@gedik.edu.tr

Gedik Vocational School, Machine and Metal Technologies Department, Istanbul Gedik University, Istanbul, Turkey

B. Li et al. (eds.), *Advances in Powder and Ceramic Materials Science 2023*, The Minerals, Metals & Materials Series, https://doi.org/10.1007/978-3-031-22622-9_14

of ZrC–TiC composite nanoparticles is of great importance. Self-propagating high temperature synthesis (SHS), which is a type of combustion synthesis, is one of the methods that has come to the fore in recent years. Although the high oxygen affinity of ZrO_2 makes it difficult to produce by the SHS method, it has been demonstrated in this study that this difficulty could be overcome with production in composite structure.

Generally, Mg reductant is used in SHS method. Aluminium is also considered suitable for some systems. In this study, the charge stoichiometry for ZrO_2–TiO_2–Mg/Al–C systems was optimized, and the maximum ZrO_2 charge stoichiometry was determined for the realization of SHS. In addition, the reduction process with Al was applied for the first time for this system, and a new chemical route was determined for the purification of the obtained product.

Experimental Study

Mg and Al were used as reductants in the synthesis of ZrC–TiC composite powder using the oxidized raw materials given in Table 1 by the SHS method, and the SHS reactions are given in Eqs. 1 and 2, respectively.

$$ZrO_2 + TiO_2 + 4\,Mg + 2C \rightarrow ZrC + TiC + 4\,MgO \tag{1}$$

$$3\,ZrO_2 + 3\,TiO_2 + 8\,Al + 2C \rightarrow ZrC + TiC + 4\,Al_2O_3 \tag{2}$$

Raw material charge stoichiometry has been optimized for SHS processes with Mg. In the SHS process with Al, stoichiometries of the reductant and charge were kept constant and compared with Mg. The experimental study plan for the SHS process is given in Table 2.

100 g of raw materials, which was mixed with a turbula mixer for 10 min and dried for 2 h, were charged into a copper crucible. Triggering was carried out by applying a voltage of 11–12 V for 4–5 s with the help of a Cr–Ni wire slightly immersed on its surface and a direct current power supply. As a result of the exothermic reaction that started at the top of the charge, the combustion wave propagating in the direction

Table 1 Purity and particle size of raw materials used in the experimental study

Raw materials	Purity, wt %	Particle size, μm
Mg	99.7	75–150
ZrO_2	99.5	<30
TiO_2	99	<75
C	98	<30
Al	99.8	75–150

Table 2 Experimental study plan for SHS process

Reductant	Reductant stoichiometry	Charge stoichiometry (for ZrC)	Specimen code
Mg	120	60–50	Mg1.2-X
Al	120	50	Al1.2–50

Table 3 Experimental study plan for chemical treatment process

Reductant	Processed specimen code	HCl concentration (M)	Product code
Mg	Mg1.2–50	12	Mg1.2–50–12
Al	Al1.2–50	12	Al1.2–50–12

of gravity spread throughout the reaction medium, and the process was completed within 15–20 s. The sponge-like product obtained was ground in agate mortar.

Chemical treatment processes were applied to remove undesired phases in the products obtained after SHS processes. For reduction with Mg, this process is HCl leaching. SHS products obtained by reduction with Al were purified by caustic fusion followed by water leaching and then HCl leaching route. The reactions taking place in the applied processes is given in Eq. 3.

$$MgO_{(s)} + 2\,HCl_{(aq)} \rightarrow MgCl_{2_{(aq)}} + H_2O_{(l)} \quad (3)$$

The experimental study plan for chemical treatment processes is given in Table 3.

The obtained products were characterized by XRD analysis. PANalytical Aeris X-Ray Powder Diffractometer (40 kV–15 mA) was used for analyses.

Results and Discussion

Self-Propagating High Temperature Synthesis Process

The results of the XRD analysis of the products obtained as a result of the SHS process performed with different charge stoichiometry are given in Fig. 1. The results showed that the SHS reaction did not occur in the charging state prepared for 60% ZrC. With the increase in the reaction enthalpy with the increasing TiO_2 charge, the combustion wave progressed and SHS was realized. In the case of using 120% of the theoretically required Mg and preparing a charge to obtain 50%TiC–50%ZrC, a significant amount of TiC and ZrC carbide was formed, although a significant amount of unreacted ZrO_2 and some TiO_2 and Mg remained in the structure. The main phase is MgO and then TiC.

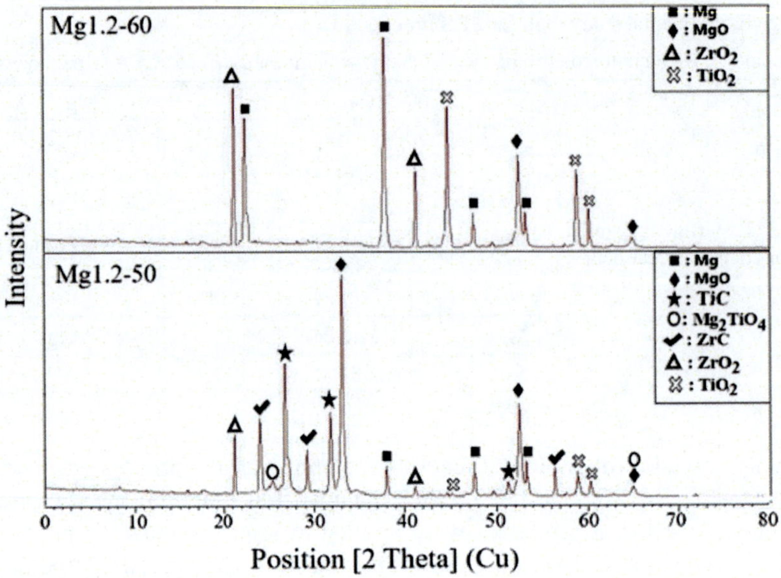

Fig. 1 XRD analysis results of SHS products of ZrO_2–TiO_2–Mg–C system for different charge stoichiometries

In the study, 120% Mg stoichiometry was used especially for lowering the amount of Mg-titanate (Mg_2TiO_4) phase which is kinetically difficult to remove in the leaching process. Also, Li et al. [7] determined the optimum Mg stoichiometry as 120% in their studies investigating the production of SHS and ZrC in the ZrO_2–Mg–C system.

The XRD analysis result of the product obtained as a result of the SHS process (ZrO_2–TiO_2–Al–C system) using Al as a reductant is given in Fig. 2 in comparison with the product obtained as a result of Mg usage. Although the main phase consists of Al, a significant amount of TiC and ZrC has been obtained. As expected, significant amount of Al_2O_3 formation and some unreacted TiO_2 and ZrO_2 formation were detected. In addition to these, it was determined that complex $ZrAlC_2$ carbide and Al_3Ti and Al_3Zr intermetallic phases were formed. This formation was also reported by Mehrizi et al. [8], and it agrees with the results obtained also by Wang et al. [9]. The formation of these carbides occurs gradually with the formation of intermediate phases [8–11].

It has been determined that the amount of ZrO_2 remained in the product was higher, and the amount of ZrC and TiC were lower as a result of reduction with Al compared to reduction with Mg.

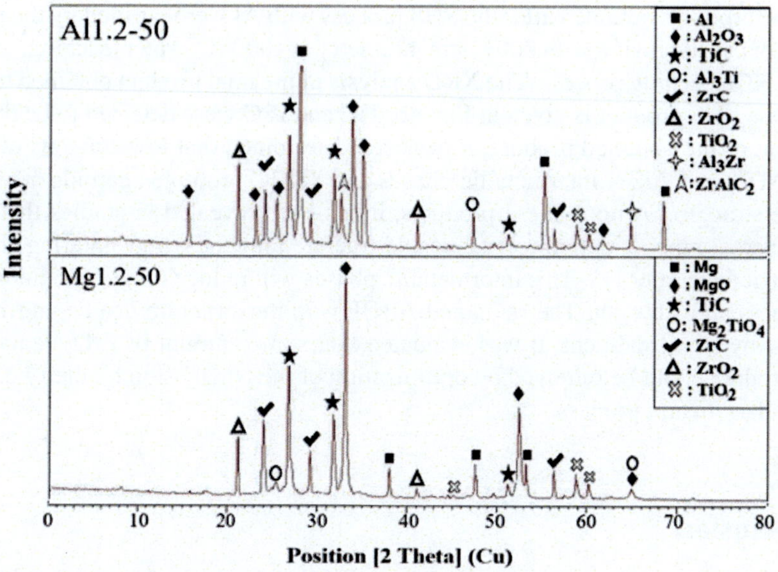

Fig. 2 XRD analysis results of SHS products for both Mg and Al usage

Chemical Treatment Process

After the SHS processes with Mg, leaching was carried out with 12 M HCl at 90 °C. XRD analysis results of the obtained product are given in Fig. 3a. The XRD analysis result of the Mg1.2–50 sample, which is the product of SHS, is also given for comparison. Accordingly, it was determined that ZrO_2 and TiO_2 could be removed almost completely, and very low amount of Mg_2TiO_4 and Mg remained in the structure. A second leaching was considered appropriate for complete purification.

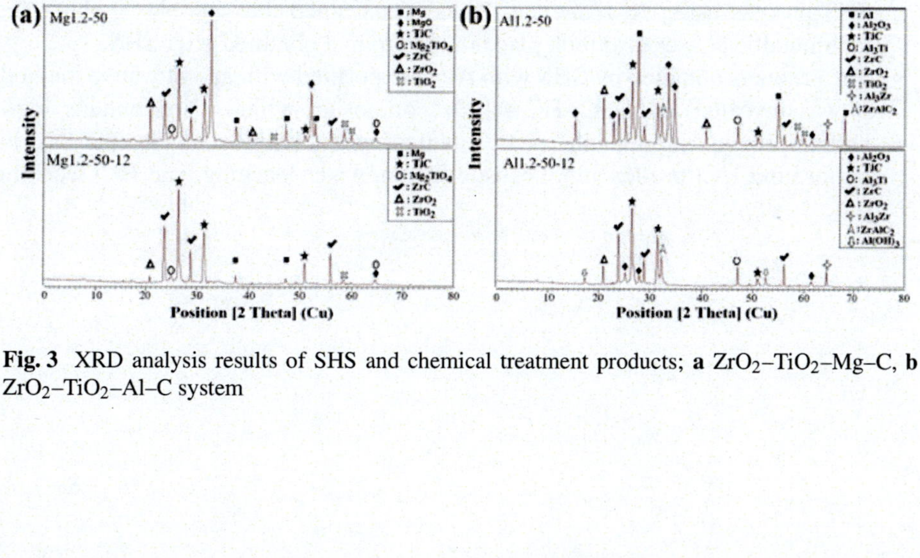

Fig. 3 XRD analysis results of SHS and chemical treatment products; **a** ZrO_2–TiO_2–Mg–C, **b** ZrO_2–TiO_2–Al–C system

The products obtained after the SHS process with Al were purified by the process route of caustic fusion with 200% stoichiometry at 1000 °C, water leaching, and then 12 M HCl leaching process. The XRD analysis of the final product obtained together with the SHS product is given in Fig. 4b. TiC and ZrC were detected to be the main content of the obtained product. It has been determined that low amounts of Al_2O_3 and $AlTi_3$ and $AlZr_3$ intermetallic phases and $ZrAlC_2$ complex carbide are present in the structure. Among these products, it has been revealed in studies that Al_2O_3 will increase the oxidation resistance and fracture toughness and sintering ability to be carried out later [12–15]. Intermetallic phases will reduce toughness but increase hardness and strength. The remained $Al(OH)_3$ in the structure can be converted to Al_2O_3 by heat treatment. It was evaluated that some amount of ZrO_2 remained in the product could be removed by optimization of the NaOH stoichiometry or by the secondary fusion process.

Conclusion

In this study, TiC–ZrC nanocomposite powder was produced with the principle of self-propagating high temperature synthesis by combustion synthesis method in ZrO_2–TiO_2–Mg/Al–C system using oxide raw materials. Significant steps of the processes using Mg and Al reductants have been optimized. Results revealed that:

- ZrO_2, which is difficult to reduce by SHS process in the case of charging alone (ZrO_2–Mg–C system), could be reduced by SHS process combined with TiO_2 (ZrO_2–TiO_2–Mg–C system).
- The charge stoichiometry has been optimized for the SHS processes carried out with Mg, and it has been determined that the maximum ZrO_2 charge should be in the stoichiometry for obtaining 50% ZrC in order for the combustion wave to progress.
- It has been demonstrated that ZrC–TiC synthesis can be performed by combined SHS process using Al reductant. It was determined that complex carbide and intermetallic phases also took place in the product obtained after SHS.
- The products obtained by SHS with Al were purified with an innovative method. It was revealed that ZrC–TiC–Al_2O_3 composite, which also contains hardness improving intermetallic phases and complex carbides, was synthesized by performing HCl pre-leaching, caustic fusion, water leaching, and HCl leaching process steps.

References

1. **Hussainova I, Voltšihhin N, Cura E, Hannula SP (2014) Densification and characterization of** spark plasma sintered ZrC–ZrO_2 composites. Mater Sci Eng, A 597:75–81
2. Landwehr SE, Hilmas GE, Fahrenholtz WG, Talmy IG (2008) Processing of ZrC–Mo cermets for high temperature applications, Part II: pressureless sintering and mechanical properties. J Am Ceram Soc 91(3):873–878
3. Barnier CBP, Thevenot F (1986) Hot-pressing kinetics of zirconium carbide. J Mater Sci 21:2547–2552
4. Umalas M, Hussainova I, Reedo V, Young DL, Cura E, Hannula SP, Lõhmus R, Lõhmus A (2015) Combined sol–gel and carbothermal synthesis of ZrC–TiC powders for composites. Mater Chem Phys 153:301–306
5. Yung DL, Kollo L, Hussainova I, Žikin A (2012) Reactive sintering of ZrC-TiC composites. Key Eng Mater 527:20–25
6. Yung DL, Hussainova I, Rodriguez M, Traksmaa R (2016) Processing of ZrC-TiC composites by SPS. Key Eng Mater 674:94–99. https://doi.org/10.4028/www.scientific.net/KEM.674.94
7. Li J, Fu ZY, Wang WM, Wang H, Lee SH, Niihara K (2010) Preparation of ZrC by self-propagating high-temperature synthesis. Ceramics Int 36(5):1681–1686. ISSN 0272–8842, https://doi.org/10.1016/j.ceramint.2010.03.013
8. Zarezadeh Mehrizin M, Beygi R, Eisaabadi G (2016) Synthesis of Al/TiC–Al_2O_3 nanocomposite by mechanical alloying and subsequent heat treatment. Ceramics Int 42:8895–8899
9. Wang Z, Liu X (2005) Reaction in the Al–ZrO_2–C system. J Mater Sci 40:4727–4735. https://doi.org/10.1007/s10853-005-0797-2
10. **Qiaodan H, Peng L, Youwei Y (2008) Microstructures, densification and mechanical properties of TiC–Al_2O_3–Al composite by field-activated combustion synthesis. Mater Sci Eng, A 486(1–**2):215–221. https://doi.org/10.1016/j.msea.2007.08.075
11. Zhao YT, Sun GX (2001) In situ synthesis of novel composites in the system Al–Zr–O. J Mater Sci Lett 20:1859–1861. https://doi.org/10.1023/A:1012889221153
12. Xu J, Zou B, Tao S, Zhang M, Cao X (2016) Fabrication and properties of Al_2O_3–TiB_2–TiC/Al metal matrix composite coatings byatmospheric plasma spraying of SHS powders. J Alloy Compd 672:251–259
13. Li Z, Wei M, Xiao K, Bai Z, Xue W, Dong C, Wei D, Li X (2019) Microhardness and wear resistance of Al_2O_3–TiB_2–TiC ceramic coatings on carbon steel fabricated by laser cladding. Ceram Int 45:115–121
14. Xiao GQ, Fu YL, Zhang ZW, Hou AD (2015) Mechanism and microstructural evolution of combustion synthesis of ZrB_2–Al_2O_3 composite powders. Ceram Int 41:5790–5797
15. Yeh C, Liu K (2022) Synthesis of TiB_2/TiC/Al_2O_3 and ZrB_2/ZrC/Al_2O_3 composites by low-exotherm thermitic combustion with PTFE activation. J Compos Sci 6(4):111. https://doi.org/10.3390/jcs6040111

Sintering Mechanism for Polycrystalline Diamond

Randal M. German

Abstract High-performance sintered diamond tools are applied to fields such as wire drawing and rock drilling. They represent a considerable advancement in hard materials sintering. Diamond particles with cobalt are liquid phase sintered to produce a dense composite using high pressure (5 GPa or more) and high temperature (1400 °C or more). Pressure is applied during heating to avoid decomposition into graphite. That pressure is amplified at grain contacts to stabilize diamond, but graphite forms at lower stress regions away from the grain contacts. Sintering occurs when melt spreads between the grains to dissolve carbon, initiating transport from graphite regions to diamond contacts. Necks nucleate on surface defects in preferred crystallographic directions. The sintered diamond exhibits properties, such as high hardness, that reflect the processing parameters or powder size, defect structure, cobalt content, peak temperature, pressure, and hold time.

Keywords Diamond · Pressure-assisted sintering · Neck growth · Liquid phase sintering · Mechanism · Phase stability

Introduction

Hard materials like diamond are difficult to consolidate. Processing temperatures are high, and an absence of plasticity disallows mechanical forming. Instead, hard composites are formed using powder techniques. At low diamond concentrations, diamonds are bonded together by sintering or hot pressing with an abundance of metal powder. At high diamond concentrations, pressure-assisted liquid phase sintering is employed using cemented carbide substrates (WC–Co). The sintering window of pressure and temperature is defined by several considerations. First, consolidation requires a pressure–temperature combination that avoids diamond graphitization. Second, the cemented carbide substrate must form a cobalt-rich liquid to infiltrate the diamonds. Diamond plasticity is another consideration. After the cycle, the sintered

R. M. German (✉)
Research Foundation Professor, San Diego State University, 282 Surfview Court, Del Mar, CA 92014, USA
e-mail: randgerman@gmail.com

B. Li et al. (eds.), *Advances in Powder and Ceramic Materials Science 2023*, The Minerals, Metals & Materials Series, https://doi.org/10.1007/978-3-031-22622-9_15

product consists of diamonds grains bonded together with interstitials consisting of cobalt with dissolved tungsten and carbon. The sintered compacts exhibit properties useful in wear, cutting, and drilling [1–3]. This article assesses the mechanism responsible for liquid phase sintering to form polycrystalline diamond (PCD).

Diamond Sintering Background

Diamond powder is irregular or polygonal depending on synthesis conditions. It packs to about 50 vol % porosity. Lower porosity is possible using bimodal or multimodal blends of large and small particles [4]. A multimode diamond particle blend might consist of 50% 17 μm, 46% 10 μm, and 4% submicrometer powders. A typical powder has a median particle size of 10–20 μm by mass or 5 μm by population. Additives might include reactive metals (e.g., Cr), nanoscale diamonds, oxygen, or nitrogen.

Below 1000 °C diamond is brittle, so the concentrated stress at particle contacts induces fracture, but then plastic deformation occurs at the sintering temperature [5–8]. The localized contact stress is up to 30-fold times the applied pressure. A threshold for particle fracture is 69.5 GPa, so a 30-fold amplification means diamond fracture starts at an applied pressure of 2.3 GPa [8, 9]. Hold times are short, so there is little grain growth. Thus, the sintered average grain size is about half the starting median particle size due to fragmentation during pressurized heating [5, 7, 8, 10]. For example, a 40 μm powder compacted at 8 GPa (no heating) results in fragments of 18–28 μm [11]. After full density is achieved, the pressure is essentially homogeneous within the diamond body.

High pressure is required during heating to avoid diamond decomposition into graphite. Sinter bonding requires sufficient temperature to induce rapid diffusion. Pure diamond sintering requires temperature in excess of 1900 °C with simultaneous pressurization to 8 GPa. To avoid such extreme conditions, cobalt is infiltrated into the diamond from a cemented carbide (WC–Co) substrate. Under pressure this melt forms near 1400 °C. Higher temperatures induce graphite formation so higher pressures follow. Mechanistically, sintering involves carbon dissolution into the liquid, preferentially from graphite, carbon diffusion in the liquid, and diamond precipitation at grain junctions. On cooling, the cobalt-rich liquid solidifies. The result is a composite of sintered diamond with cobalt-rich interstitials, as captured in Fig. 1. This form of liquid phase sintering is sensitive to diffusion distances, so it is typically applied to particles smaller than about 10 μm.

Once liquid solvent metal infiltrates the diamond, then bonding occurs by solution reprecipitation. Graphite is the preferred carbon source and diamond is the preferred growth product. Carbon diffuses from lower density graphite regions (≈2.3 g/cm^3) to precipitate as higher density diamond (≈3.5 g/cm^3) at grain contacts. Cobalt-rich interstitials fill between the diamond grains. After cooling the microstructure consisting of about 90 wt % diamond, 8 wt % cobalt, and 2 wt % tungsten. The enabling technology is termed high-pressure high-temperature (HPHT) sintering. It

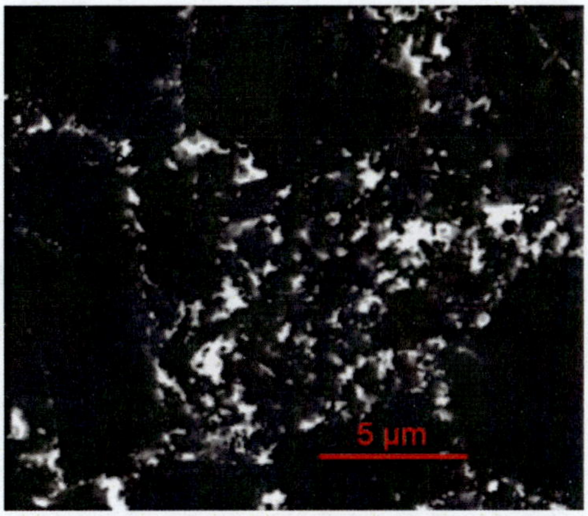

Fig. 1 Cross-sectional scanning electron micrograph showing sinter bonded diamond (black) with surrounding cobalt-rich interstitials (white)

is sensitive to a host of processing factors. The sintering kinetics depend on the rate of liquid formation and infiltration, diffusion in the liquid, and precipitation at grain junctions. The sintered structure has considerable strength, hardness, and toughness, reaching 1500 MPa, 7700 HV, and 12 MPa$\sqrt{}$m. Effective sintering depends on key processing factors of particle size, cobalt content, peak temperature, pressurization rate, peak pressure, and hold time.

Sintering Window

Graphite is the stable form of carbon at one atmosphere pressure and room temperature but remains metastable up to at least 800 °C. Slow heating to intermediate temperatures converts diamond to graphite. To circumvent decomposition, diamond sintering is performed at temperature–pressure combinations where diamond is stable using HPHT devices [1, 2]. As plotted in Fig. 2, the pressure–temperature (*P-T*) trade-off between graphite and diamond is as follows (in GPa and °C):

$$T = 400P - 776 \text{ or } T = 370P - 536$$

This is the equilibrium condition, but metastable diamond survives brief times even when heated to 800–1000 °C [12–14]. Metals such as Co, Fe, and Ni accelerate both the formation of diamond from graphite and the reversion of diamond to graphite. Hence, Co infiltration after heating and pressurization avoids excess graphitization. As shown in Fig. 2, there is a metastability gap. Sufficient pressure is needed to cross into the diamond stable region prior to reaching the peak temperature. Short hold times avoid grain growth.

A further consideration is diamond softening. The brittle–ductile boundary is included in Fig. 2. Another constraint is defined by the *P-T* combinations that produce solvent metal melting and infiltration. This occurs at the Co–C eutectic temperature.

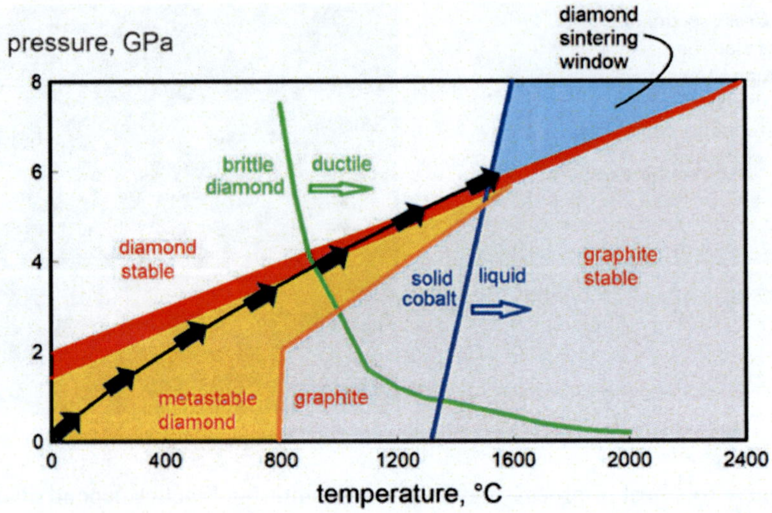

Fig. 2 Diamond sintering temperature and pressure plot showing the boundaries that explain events during heating and pressurization. The region of diamond metastability is added to reconcile the temperature–pressure pathway (black arrows) used to sinter diamond

At one atmosphere pressure, this is at 1320 °C. Under HPHT conditions, the eutectic temperature is slightly higher since pressure stabilizes solid cobalt, requiring about 1400 °C for 5.5 GPa [1, 15]. The carbon solubility jumps when liquid first forms, allowing dissolution of graphite islands formed during heating. Estimates on liquid metal infiltration give the capillary attraction into the diamond at about 2 MPa (10% porosity, 10 μm diamond grains, 1.5 μm pores, 0.7 J/m^2 surface energy). The melt viscosity for cobalt at 7 GPa pressure and 1500 °C is 27 mPa s, giving spontaneous infiltration (less than 1 s). Experiments report 3 mm fill depth in 10–20 s [16]. Thus, heat transfer and melting are the controlling steps not infiltration [17, 18]. At HPHT temperatures, cobalt diffusivity is $3 \cdot 10^{-9}$ m^2/s. So, only short times are required for bonding. The absence of a post-sintering cobalt concentration gradient is evidence of rapid diffusion and sintering [19].

The sintering window is based on diamond stability, metastability, melt formation, and plasticity. Short times at temperature are sufficient to induce the solution–reprecipitation events responsible for diamond bonding. Based on several studies, typical consolidation conditions are 7 μm diamond powder, substrate with 15% cobalt (WC-15Co), 6.8 GPa pressure, 1530 °C peak temperature with a 6 min hold [5, 20, 21].

Solution–Reprecipitation Mechanism

Stress is lowest between the diamond contact points. Heat reverts these low-pressure regions into graphite. The stress variation around a grain gives a mixture of surface graphite and diamond when the matrix liquid forms. Diamond is insoluble in cobalt at 1000 °C, but reaches 141 ppm at 1300 °C [22]. On the other hand, graphite is soluble in cobalt, reaching about 1 at % at 1000 °C. The graphite solubility jumps to 11.6 at % at the eutectic temperature. This is key to the solution–reprecipitation mechanism. That melt preferentially dissolves the graphite and precipitates diamond [17]. Accordingly, cobalt converts hexagonal-close packed graphite into cubic diamond. Times as short as 10 to 30 s at the peak pressure and temperature are satisfactory to form sinter bonds. Sinter bonds are evident growing from surface defects, such as screw dislocation spirals, as nucleation sites, by solution–reprecipitation. The associated ledges allow for atoms to deposit along preferred planes in preferred directions.

The sintering rate increases with temperature, following an Arrhenius temperature dependence. For temperature and pressure in the diamond stable region, say 1600 °C and 6 GPa, the measured and calculated carbon diffusivity in the melt is $2{\cdot}10^{-9}$ to $5{\cdot}10^{-9}$ m^2/s. The activation energy for carbon diffusion in liquid cobalt is estimated at 145 kJ/mol. At 1500 °C, cobalt atoms move more than 0.5 μm per second, based on more than 1200 position changes. Higher temperatures induce faster sintering. Accordingly, the characteristic diffusion distance is 200 μm in 10 s. That atomic motion involves heterogeneous surface defects [19]. The diffusion distance is a fraction of the particle size as sketched in Fig. 3 because regions of high-pressure particle contact are close to low-pressure graphite regions. Diffusion in the liquid makes transport comparatively rapid, reaching 0.5 μm/s.

Bonds grow from surface nucleation sites by spiral growth around screw dislocations. Figure 4 shows the idea. The screw dislocation ledge expands by atom additions resulting in horizontal extension and upward growth. This growth mechanism is observed in liquid-phase sintering of other anisotropic surface energy crystals,

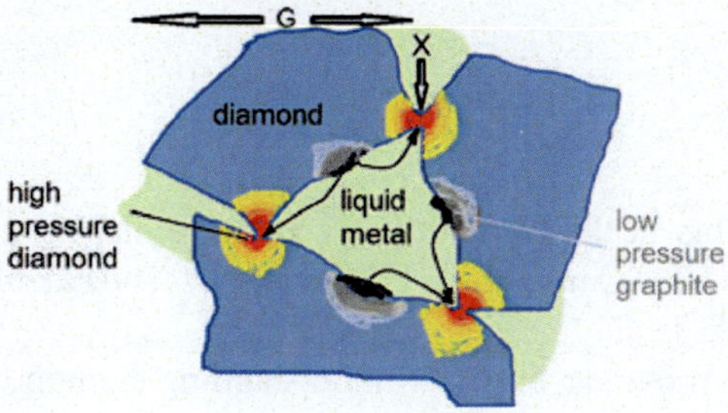

Fig. 3 Sketch of how particle contacts under high-pressure stabilize diamond while neighboring regions under lower pressure provide graphite. The transport of dissolved carbon in the liquid metal results in sinter bonds at the diamond particle contacts. The grain size is G and interparticle neck size is X

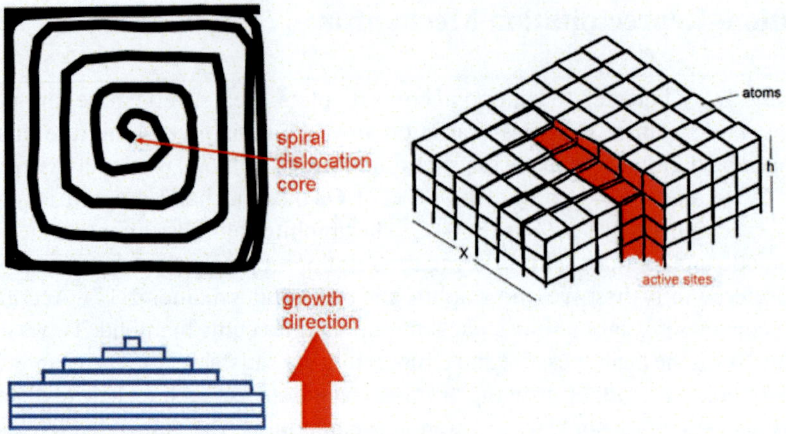

Fig. 4 Left drawing shows two views of a spiral growth pattern around a screw dislocation. The drawing illustrates the growth steps formed by adding atoms to the edge of the spiral. The plateaus expand outward while the crystal grows upward from a screw dislocation or other surface defects

such as WC in Co [23]. It is the limited solvation and deposition sites along with the limited solubility that determines the sintering rate [24].

To model bond growth, examine the bond size *X* for a grain size *G*, as sketched in Fig. 3. After liquid infiltration, carbon dissolves at lower pressure graphite-rich zones and precipitates as necks between diamond grains. The precipitation rate is limited by availability of precipitation sites.

Neck size depends on transport from surfaces producing graphite to diamond sinter bonds. Because of the limited additive sites for the spiral growth, the height h is approximately half the width X. The total number of atomic sites for this geometry is 3 X^2/a^2 where a is the atomic diameter. The active sites are at the ledge with a population of 2 X/a. Growth also depends on the carbon solubility in the matrix C (set to 0.2) and the fraction of atom sites providing growth. Accounting for the ledge area, arrival sites, and geometry gives

$$\frac{X}{G} = \left[\frac{gtD_v\gamma\Omega CA_F}{kTG^M}\right]^{1/N}$$

Figure 5 plots this neck size ratio X/G versus sintering hold time for 10 μm diamond grains at 1500 °C in a liquid cobalt matrix. The calculation parameters are given below in Table 1.

The calculations provide a basis for understanding diamond sintering. Due to the high diffusivity of carbon in liquid cobalt, bond growth is rapid, but slows after the first few minutes due to limited deposition sites. After 10 min the predicted neck size ratio X/G is 0.146, and after 60 min, it is 0.214; 70% of the neck growth occurs in the first 17% of the hold time. Short times provide considerable sinter bonding as evident in Fig. 6, a high magnification electron backscatter diffraction image. Some

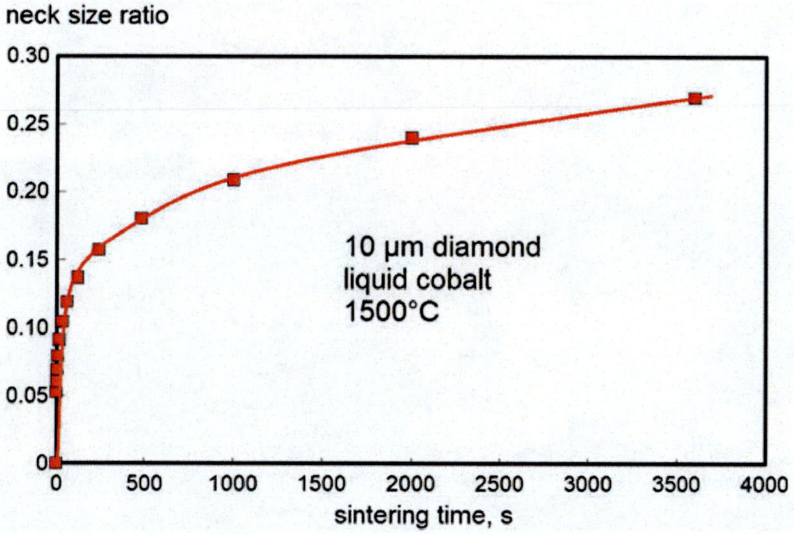

Fig. 5 Neck size ratio as a function of sintering time for diamond starting with 10 μm grains

Table 1 Calculation parameters for estimating the neck growth of diamond grains in a liquid cobalt matrix; t = time in seconds, k = gas constant (8.31 J/mol K)

Parameter	Selected value
a atomic diameter, m	$1.54 \cdot 10^{-10}$
γ surface energy, J/m^2	0.60
Ω atomic volume, m^3	$5.68 \cdot 10^{-30}$
C solubility, atom fraction	0.2
D_V diffusivity, m^2/s	$4 \cdot 10^{-9}$
G grain size, m	10^{-5}
T temperature, K	1773
M size scale term	3
N mechanistic constant	5
g geometric term	80
A_F area fraction	a/3X

sinter bonds are circled in red, showing grain contacts and grain boundaries where the crystals intersect. The HPHT apparatus is expensive, so longer hold times decrease productivity, increase cost, but give only a small improvement in bonding.

The diamond grains bond together with the remainder of the grain perimeter taken up by solidified matrix. Prolonged heating can induce grain growth, with elimination of the small grains and growth of the large grains. Simultaneously, the cobalt liquid pools coarsen. This can degrade properties. Because surface energies determine the degree of grain bounding, via the dihedral angle, where the terminal neck size is set by surface energies, long hold times give little gain in the relative degree of bonding.

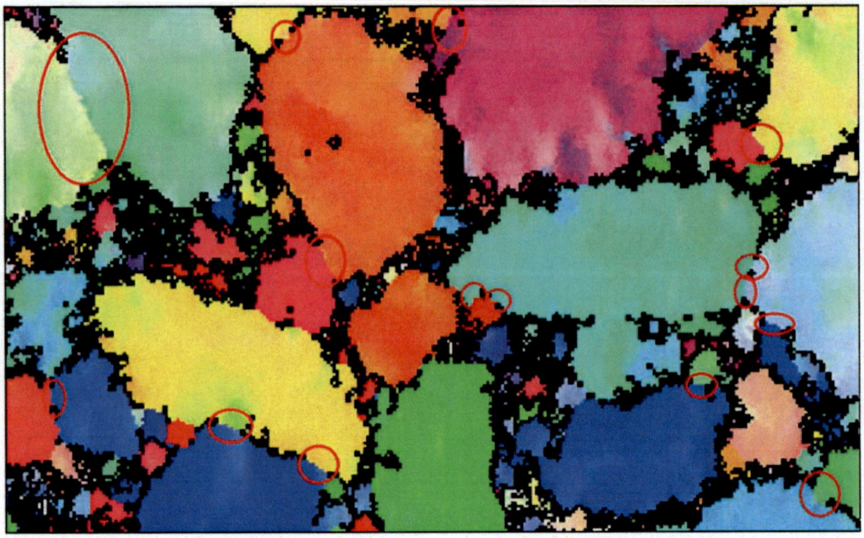

Fig. 6 Electron backscatter diffraction image of diamond grains where crystal orientation is colorized and diamond-diamond sinter bonds are identified

Summary

Sintered diamond requires HPHT processing as defined by the sintering window. For a 12 μm powder, sintering without a solvent metal requires 7 GPa and 2130 °C to reach full density. The infiltration of a solvent metal from a WC-13Co cemented carbide substrate relaxes the pressure–temperature demands to 5.5 GPa and 1450 °C for the same 12 μm powder. This agrees with the sintering window isolated here, as evident in Fig. 7. After consolidation, the wt % cobalt content in the diamond is about half that of the substrate. Bonding follows a model for liquid phase sintering with limited interface sites, such as associated with surface growth spirals.

Performance attributes for PCD follow from the sintering behavior, where full density is an important first criteria. The HPHT devices are expensive, so productivity is optimized using short holds in the HPHT sintering window. Grain size is smaller than the particle size due to particle fracture during heating.

The model can be verified by heating diamond powder compacts to various points in the consolidation cycle, and then rapidly cooling to room temperature. Observations on the microstructure would include measuring composition gradients, solvation or reaction events, and degree of bonding. Changes in bonding, phases, and properties would verify the sequence of steps making up this sintering trajectory.

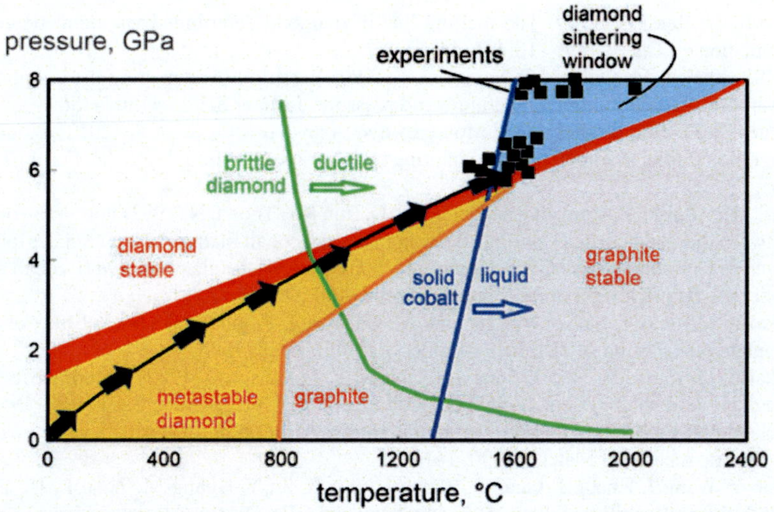

Fig. 7 Demarcation of the sintering window with several full density experiments, typically using cobalt as the solvent metal

References

1. Katzman H, Libby WF (1971) Sintered diamond compacts with a cobalt binder. Science 172:1132–1134
2. Belnap JD (2010) Sintering of ultrahard materials. In: Fang ZZ (ed) Sintering of advanced materials. Woodhead, Oxford, pp 389–414
3. Belnap JD, Griffo A (2004) Homogeneous and structured PCD/WC-Co materials for drilling. Diamond Related Mater 13:1914–1922
4. Hong SM, Akaishi M, Kanda H, Osawa T, Yamoaka S (1991) Dissolution behavior of fine particles of diamond under high pressure sintering conditions. J Mater Sci Lett 10:164–166
5. DeVries RC (1975) Plastic deformation and "Work-Hardening" of diamond. Mater Res Bull 10:1193–1200
6. Gasc J, Wang Y, Yu T, Benea IC, Rosczyk BR, Shinmei T, Irifune T (2015) High-pressure, high-temperature plastic deformation of sintered diamonds. Diamond Related Mater 59:95–103
7. Uehara K, Yamaya S (1990) High pressure sintering of diamond by cobalt infiltration. In: Saito S, Fukunaga O, Yoshikawa M (eds) Proceedings 1st international conference on new diamond science and technology, pp 203–209
8. Guan S, Peng F, Liang H, Fan C, Tan I, Wang Z, Zhang Y, Zhang J, Yu H, He D (2018) Fragmentation and stress diversification in diamond powder under high pressure. J Appl Phys 134:215902
9. Ngwekhulu TT (2018) Effect of Diamond Grain Size on Magnetic Properties of the Cobalt Phase in PCD Compacts, MS Thesis, University of the Witwatersrand, Johannesburg
10. Brookes CA, Brookes EJ (1991) Diamond in perspective: a review of mechanical properties of natural diamond. Diamond Related Mater 1:13–17
11. Bobrovnitchii GS, Osipov OS, Filgueira M (2003) Some peculiarities of the diamond micro-powder sintering. Inter J Refract Met Hard Mater 21:251–258
12. Qian J, Pantea C, Voronin G, Zerda TW (2001) Partial graphitization of diamond crystals under high-pressure and high-temperature conditions. J Appl Phys 90:1632–1637
13. Davies G, Evans T (1972) Graphitization of diamond at zero pressure and at a high pressure. Proc Royal Soc Lond A 328:413–427

14. Glosli JN, Ree FH (1999) The melting line of diamond determined via atomistic computer simulations. J Chem Phys 110:441–446
15. Li SJ, Akaishi M, Ohsawa T, Yamaoka S (1990) Sintering behaviour of the diamond-super invar alloy system at high temperature and pressure. J Mater Sci 25:4150–4156
16. Skury ALD, Bobrovinichii GS, Monteiro SN, Azevdeo MG, Silva A (2012) Liquid phase migration during diamond powder sintering by infiltration method. Mater Sci Forum 727:450–455
17. Park JK, Akaishi M, Yamaoka M, Fukunaga O, Eun KY, Yoon DN (1992) Formation of bridges between diamond particles during sintering in molten cobalt matrix. J Mater Sci 27:4695–4697
18. Lima FTC, Bobrovnitchii GS, Filgueira M (2005) Study of the diamond 5 wt % cobalt sintering under the HPHT lowest limit. Mater Sci Forum 498:225–230
19. Akaishi M, Sato Y, Setaka N, Tsutsumi M, Ohsawa T, Fujunaga O (1983) Effect of additive graphite on sintering of diamond. Am Cer Soc Bull 62:689–695
20. Akaishi M, Kanda H, Sato Y, Setaka N, Ohsawa T, Fukunaga O (1982) Sintering behaviour of the diamond-cobalt system at high temperature and pressure. J Mater Sci 17:193–198
21. Boland JN, Li XS (2010) Microstructural characterisation and wear behaviour of diamond composite materials. Mater 3:1390–1419
22. Tian Y, Wang J, Zhang J, Guan S, Zhang L, Wu B, Su Y, Huang M, Zhou L, He D (2020) Solubility and stability of diamond in cobalt under 5 GPa. Diamond Related Mater 110:108158
23. German RM (1996) Sintering theory and practice. Wiley-Interscience, New York
24. German RM (2014) Sintering: from empirical observations to scientific principles. Elsevier, Oxford

Combustion Synthesis of B_4C–TiB_2 Composite Nanoparticle by Self-Propagating High-Temperature Synthesis (SHS) in B_2O_3–TiO_2–Mg–C System

Ozan Coban, Mehmet Bugdayci, Serkan Baslayici, and M. Ercan Acma

Abstract In this study, B_4C-TiB_2 nanocomposite powder was synthesized from oxide raw materials with the principle of magnesiothermic reduction in B_2O_3–TiO_2–Mg–C system by SHS method. For the SHS process, Mg and C stoichiometries were optimized with thermochemical simulation, and composite charge stoichiometry and Mg particle size were optimized with XRD, BET and SEM analyzes. Optimization of acid concentration, leaching temperature, and leaching time parameters has been provided for the HCl leaching processes carried out to remove undesired by-products after SHS. In addition, pH and temperature changes during leaching were analyzed and an innovative application of modified leaching with H_2O_2 and carbonic acid addition was investigated. The results showed that by optimizing the process steps for the synthesis of B_4C–TiB_2 composite nanoparticle by the SHS method, a commercial grade product with a surface area of 30.6 m^2/g, and a particle size of 193 nm was obtained.

Keywords Boron carbide · Titanium diboride · Composite powder · Nanoparticle synthesis · Advanced ceramics

---

O. Coban (✉) · M. E. Acma
Metallurgical and Materials Engineering Department, Istanbul Technical University, Istanbul, Turkey
e-mail: ozan.coban@gedik.edu.tr

O. Coban
Gedik Vocational School, Machine and Metal Technologies Department, Istanbul Gedik University, Istanbul, Turkey

M. Bugdayci
Chemical Engineering Department, Yalova University, 77200 Yalova, Turkey

S. Baslayici
Vocational School, Construction Technology Department, Istanbul Medipol University, Istanbul, Turkey

B. Li et al. (eds.), *Advances in Powder and Ceramic Materials Science 2023*, The Minerals, Metals & Materials Series, https://doi.org/10.1007/978-3-031-22622-9_16

Introduction

It is of great importance to synthesize composite powders that combine the properties of advanced technology ceramics with superior mechanical, physical, and chemical properties. Boron Carbide (B_4C) has very high hardness and wear resistance and high temperature resistance. Thanks to its superior properties, it has wide usage areas as armor technologies, hard coatings, nuclear applications, and high-temperature semiconductor [1–9]. Beside, titanium diboride (TiB_2) has similar superior properties, and it stands out with its features such as high fracture toughness and thermal shock resistance [10–13]. The low plasticity of B_4C and the oxide layer formed on its surface reduce its sintering ability and its low toughness limits its use. Studies have shown that these disadvantages can be eliminated as a result of its combination with TiB_2 [14–16].

There are various methods for synthesizing B_4C and TiB_2 powders [17–23]. In general, composite production is applied by sintering together by producing separately, not in composite structure. However, combustion synthesis both provides a single-stage dust synthesis in the composite structure, and it eliminates the disadvantages such as high temperature and/or pressure and extensive mechanical milling in other methods. Recently, there are some studies on the production of B_4C–TiB_2 with SHS. Nikzad et al. [24] produced using elemental raw materials, mechanical activation, and chemical boosters, but these make the method expensive. Bahabad et al. [25], on the other hand, produced by combustion synthesis using oxidized raw materials, but triggering was provided under continuous argon flow at 900 °C in volume combustion mode. This increases the cost too much. The grain size of the raw materials used is another important parameter that needs to be optimized. While a high particle size reduces the reaction interface, an extremely low particle size may cause a decrease in efficiency due to evaporation and scattering, while increasing the speed of the combustion wave increases the combustion temperature. There are studies examining the effect of particle size on the synthesis of high-tech ceramic powders with SHS [26–29]. However, this parameter should be optimized for each system.

Experimental Study

In this study, B_4C–TiB_2 composite powders were produced with the principle of magnesiothermic reduction in the B_2O_3–TiO_2–Mg–C system by self-propagating high temperature synthesis (SHS) method. The reaction given in Eq. 1 explains the basic principle of production.

$$TiO_2 + 3\,B_2O_3 + 11\,Mg + C \rightarrow TiB_2 + B_4C + 11\,MgO \quad (1)$$

Table 1 Purity and particle size of raw materials used in the experimental study

Raw materials	Purity, wt %	Particle size, μm
Mg	99.7	150–250/75–150/10–50
B_2O_3	96.5	<55
TiO_2	99	<75
C	98	<30

In the first stage, thermochemical simulation was performed using Factsage software. Stoichiometric optimization was performed for magnesium used as a reductant and carbon used as a carbide former. Then, the effect of the grain size of the reductant on the phase amounts in the SHS product was investigated by using magnesium with three different grain sizes. The purity and particle size of raw materials used in the experimental study are given in Table 1.

SHS processes were carried out in the atmospheric environment under the determined optimum conditions. According to the thermochemical simulation result given in Fig. 1, the adiabatic temperature and predicted phase amounts were evaluated and the optimum charge stoichiometry was determined as TiO_2:B_2O_3:Mg:C = 1:3:12:1.6. That is, the optimum charge percentages for Mg and C were determined as 110 and 160%, respectively.

HCl acid leaching processes were carried out to remove the undesired phases in the product obtained by SHS. Acid concentration, leaching temperature, leaching time, and cascading leaching parameters were optimized. Leaching conditions are given in Table 2.

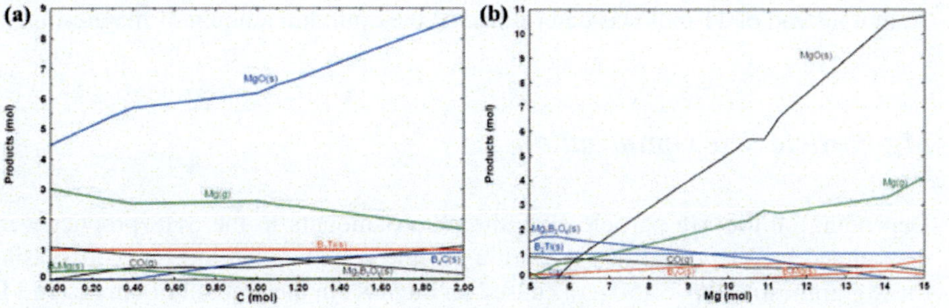

Fig. 1 Termochemical simulation results for varying **a** carbon, **b** magnesium moles

Table 2 Conditions of leaching experiments for each sample

Acid concentration	Leaching temperature (°C)	Leaching duration (min)	Stirring rate	Solid/liquid ratio
6.5 M/8.5 M/10.5 M/12.5 M	25/40/65/75/90	30/45/60/75	500 rpm	1/5
Second Leach (10.5 M HCl)	90 °C	60 min		

100 g of raw material, which was mixed with a turbula mixer for 10 min and dried for 2 h, was charged into a copper crucible. Triggering was carried out by applying a voltage of 11–12 V for 4–5 s with the help of a Cr–Ni wire slightly immersed on its surface and a direct current power supply. As a result of the exothermic reaction that started at the top of the charge, the combustion wave propagating in the direction of gravity spread throughout the reaction medium, and the process was completed within 15–20 s. The spongy product obtained was ground in agate mortar. Leaching processes were carried out using a heated magnetic stirrer. The obtained products were characterized by XRD, BET, and SEM analysis. PANalytical Aeris X-ray powder diffractometer (40 kV–15 mA), micromeritics ASAP 2020 surface area and porosity analyzer, and Zeiss GeminiSEM 500 field emission scanning electron microscope were used for analysis.

Results and Discussion

Thermochemical Simulation Results

The predicted phase amounts were evaluated according to the changing carbon and magnesium mole ratios by thermochemical simulation, and the result is given in Fig. 1. Accordingly, 1.6 mol instead of 1 mol was determined as the optimum carbon amount due to the 44% decrease in the Mg-borate (Mg_3B_2O6) phase, which is difficult to remove in the subsequent leaching process, and the 21% increase in the B_4C amount. Similarly, due to the significant decrease in Mg-borate ($Mg_3B_2O_6$) phase, 12 mol instead of 11 mol was determined as the optimum amount of magnesium.

Mg Particle Size Optimization

Depending on the Mg particle size, the phase amounts in the SHS product were determined by XRD analysis. According to the results given in Fig. 2, increasing the Mg grain size above 150 μm caused an increase in the amount of unreacted Mg in the product due to the reduction of the reaction interface. As a result of the Mg grain size being below 50 μm, although the amount of Mg decreased and the amount of MgO increased, it caused a decrease in the amount of B_4C–TiB_2, which is the desired phases. This is an indication that a significant amount of Mg does not enter the reduction reaction and reacts with the oxygen in the environment to form MgO. The reason for this is that the extremely fine-grained magnesium evaporated with the effect of the temperature, resulting from the reaction and is scattered under the pressure effect. According to the results obtained, the optimum Mg particle size was determined as 75–150 μm.

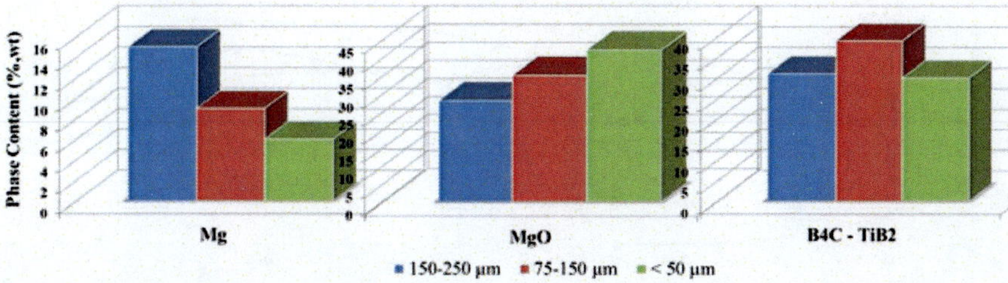

Fig. 2 Phase contents of SHS product for varying Mg particle sizes

Acid Concentration Optimization

The products obtained after SHS were leached with different concentrations of HCl. The XRD results of the obtained products are given in Fig. 3. Although the main product is TiB_2, B_4C can also be obtained. It was determined that TiO_2 and Mg from the undesired phases could not be completely removed. It was determined that the increase of acid concentration decreased, especially the Mg amount, and it could be largely removed at 10.5 M acid concentration. Further increase in concentration did not have a significant effect. Obtained results showed that 10.5 M is the optimum acid concentration, and secondary leaching is required to obtain high purity product. The obtained result was found to be in agreement with studies that optimized the acid concentration for SHS production of B_4C and TiB_2 separately [31, 32].

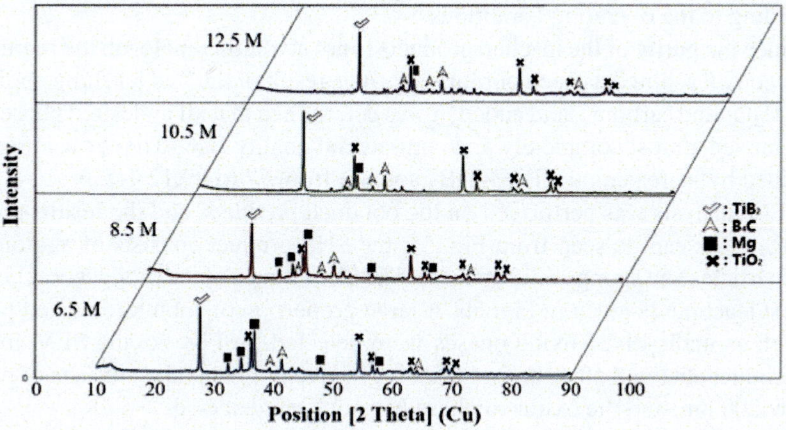

Fig. 3 XRD analysis results of leached products

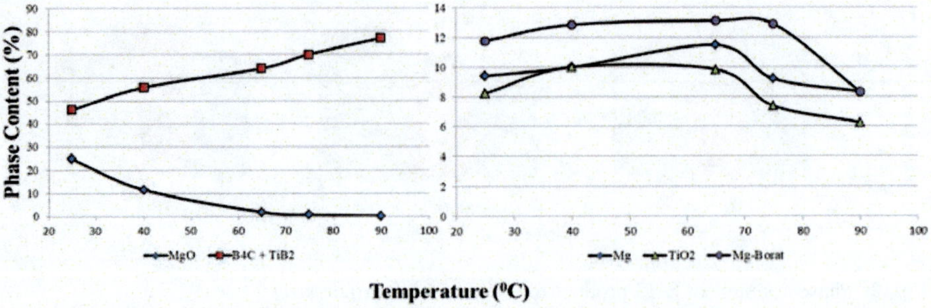

Fig. 4 Effect of leaching temperature on phase contents of leached product

Optimization of Leaching Temperature

The phase amounts were determined as a result of the XRD analysis of the products obtained after the leaching process carried out at different temperatures, and the obtained results are given in Fig. 4. With the increase of the leaching temperature, the amount of MgO decreased, while the amount of B4C–TiB_2 increased accordingly. It has been determined that the undesired Mg, TiO_2 and Mg-borate phases can be removed after MgO is almost completely removed above 60 °C. It has been determined that Mg-borate phases, which are especially difficult to remove, can be removed above 75 °C. Ipekci et al. [32] agreed with the result obtained. For this reason, while the minimum temperature to be applied was determined as 75 °C, it was evaluated that 90 °C was the optimum temperature, although it would vary according to the operating conditions.

Since the purity of the leached products is not at a sufficient level, the requirement for second leaching has been put forward. As a result of the 2nd leaching application with H_2O_2 and carbonic acid added, it was determined that all undesired phases could be removed almost completely and commercial quality B_4C–TiB_2 powder could be obtained by increasing the B_4C–TiB_2 amount from 77 to 99.12%.

SEM analysis was performed on the obtained products, and the results are given in Fig. 5. As can be seen from Fig. 5a, the SHS product consists of agglomerated nanoparticles and spongy oxide phases. The SEM image of the product obtained after the 1st leaching is given in Fig. 5b. A large proportion of submicron-sized powders as well as undissolved oxide phases were detected. In Fig. 5c, the SEM image of the product obtained after the 2nd leaching is given. It generally consists of particles below 200 nm, and the oxidized phase has not been detected.

BET analysis was also performed on the products obtained and it was determined that nanoparticle synthesis with a particle size of 193.5 nm could be performed.

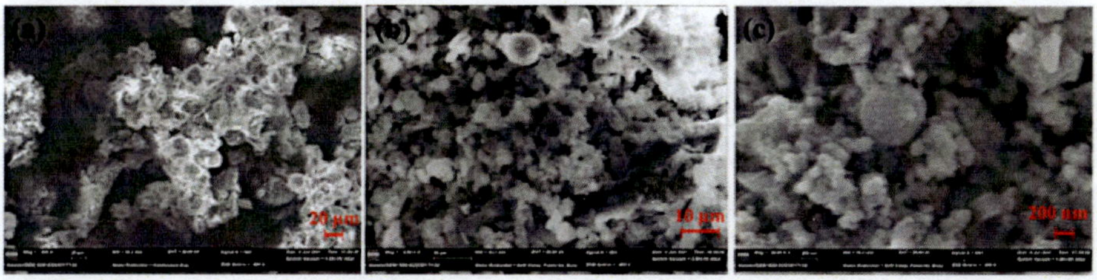

Fig. 5 SEM micrographs of **a** SHS product (Mag. 500X), **b** 1st leach product (Mag. 2500X), **c** 2nd leach product (Mag = 50.000X)

Conclusion

In this study, B_4C–TiB_2 composite nanoparticles were synthesized under atmospheric conditions by combustion synthesis method using oxidized raw materials, magnesium reductant and carbon. The optimum stoichiometry of Mg and C for the SHS process was investigated thermochemically and the optimum stoichiometry was determined as TiO_2:B_2O_3:Mg:C = 1:3:12:1.6. The Mg particle size was optimized and it was found that below 75 μm, the yield decreased due to scattering and evaporation. The optimum Mg particle size was determined as 75–150 μm. Acid concentration, leaching temperature and leaching time parameters were optimized for the leaching process carried out to remove undesired products after SHS, and were determined as 10.5 M, 90 °C and 60 min, respectively. Especially, the high temperature requirement has been put forward for the removal of Mg-borate phases. It was revealed that the purity, which was 77% as a result of the single-stage leaching application, could be increased to 99.12% as a result of the second leaching application. The leaching process was also found to reduce the particle size to around 200 nm. As a result, commercial quality and very high surface area nanoparticle synthesis was carried out and the optimization of the production processes was carried out with SHS in powder extraction for industrial production of B_4C–TiB_2 composite.

References

1. Dariel MP, Frage N (2012) Reaction bonded boron carbide: recent developments. Adv Appl Ceram 111(5–6):301–310. https://doi.org/10.1179/1743676111Y.0000000078
2. Xin L, Minjun L, Shuaibo G, Shu Y, Xiaofeng W, Pengfei X (2019) Effect of initial compositions on boron carbide synthesis and corresponding growth mechanism. Adv Appl Ceram 118(8):442–450. https://doi.org/10.1080/17436753.2019.1664792
3. Yaşar ZA, Haber RA (2021) Effect of sintering temperature and applied pressure on the properties of boron carbide-silicon carbide composites. J Superhard Mater 43:392–404. https://doi.org/10.3103/S1063457621060022
4. Crouch IG, Eu B (2017) Ballistic testing methodologies, vol. 11 of the science of armour materials. Woodhead Publishing, pp 639–673. https://doi.org/10.1016/B978-0-08-100704-4.00011-6

5. Domnich V, Reynaud S, Haber R, Chhowalla M (2011) Boron carbide: structure, properties and stability under stress. J Am Ceram Soc 64(11):3605–3628. https://doi.org/10.1111/j.1551-2916.2011.04865.x
6. Savio SG, Sambasiva Rao A, Rama Subba Reddy P, Madhu V (2019) Microstructure and ballistic performance of hot pressed and reaction bonded boron carbides against an armour piercing projectile. Adv Appl Ceram 118(5):264–273. https://doi.org/10.1080/17436753.2018.1564416
7. Wood C (1988) Materials for thermoelectric energy conversion. Rep Prog Phys 51(4):459. https://doi.org/10.1088/0034-4885/51/4/001
8. Suri AK, Subramanian C, Sonber JK, Ch MTSR (2010) Synthesis and consolidation of boron carbide: a review. Int Mater Rev 55(1):4–40. https://doi.org/10.1179/095066009X12506721665211
9. Sauerschnig P, Watts JL, Vaney JB, Talbot PC, Alarco JA, Mackinnon IDR et al (2020) Thermoelectric properties of phase pure boron carbide prepared by a solution-based method. Adv Appl Ceram 119(2):97–106. https://doi.org/10.1080/17436753.2019.1705017
10. Mukhopadhyay A, Raju GB, Basu B, Suri AK (2009) Correlation between phase evolution, mechanical properties and instrumented indentation response of TiB_2-based ceramics. J Euro Ceramics Soc 29(3):505–516. https://doi.org/10.1016/j.jeurceramsoc.2008.06.030
11. Ricceri R, Matteazzi P (2004) A fast and low-cost room temperature process for TiB_2 formation by mechanosynthesis. Mater Sci Eng, A 379(1–2):341–346. https://doi.org/10.1016/j.msea.2004.02.064
12. **Górnya G, Rączkaa M, Stobierskia L,** Rożniatowskib K, Rutkowskia P (2009) Ceramic composite Ti_3SiC_2–TiB_2-Microstructure and mechanical properties. Mater Charact 60(10):1168–1174. https://doi.org/10.1016/j.matchar.2009.03.011
13. Bilgi E, Camurlu HE, Akgun B, Topkaya Y, Sevinc N (2008) Formation of TiB2 by volume combustion and mechanochemical process. Mater Res Bull 43(4):873–881. https://doi.org/10.1016/j.materresbull.2007.05.001
14. Madhav Reddy K, Guo JJ, Shinoda Y, Fujita T, Hirata A, Singh JP et al (2012) Enhanced mechanical properties of nanocrystalline boron carbide by nanoporosity and interface phases. Nat Commun 3:1052. https://doi.org/10.1038/ncomms2047
15. Heydari MS, Baharvandi HR (2015) Comparing the effects of different sintering methods for ceramics on the physical and mechanical properties of B_4C–TiB_2 nanocomposites. Int J Refract Metal Hard Mater 51:224–232. https://doi.org/10.1016/j.ijrmhm.2015.04.003
16. Ivzhenko VV, Kryl' AO, Kryl' YA et al (2014) Study of aeroabrasive wear of hot-pressed materials of the B4C–TiB_2 system. J Superhard Mater 36:187–192. https://doi.org/10.3103/S106345761403006X
17. Munir ZA, Anselmi-Tamburini U (1989) Self-propagating exothermic reactions: the synthesis of high-temperature materials by combustion. Mater Sci Rep 3(7–8):277–365. https://doi.org/10.1016/0920-2307(89)90001-7
18. Xue H, Munir ZA (1996) Extending the compositional limit of combustion synthesized B_4C–TiB_2 **composites by field activation. Metall Mater Trans B 27:475–480**
19. Fard HSP, Baharvandi HR, Abdizadeh H, Shahbahrami B (2008) Chemical synthesis of nano-titanium diboride powders by borothermic reduction. Int J Mod Phys B 22(18–19):3179–3184. https://doi.org/10.1142/S0217979208048085
20. Pei LZ, Xiao HN (2009) B_4C–TiB_2 composite powders prepared by carbothermal reduction method. J Mater Process Technol 209(4):2122–2127. https://doi.org/10.1016/j.jmatprotec.2008.05.003
21. Cakir E, Ergun C, Sahin FC, Erden I (2010) In situ synthesis of B4C/TiB2 composites from low cost sugar based precursor. Defect Diffusion Forum 297–301:52–56. https://doi.org/10.4028/www.scientific.net/DDF.297-301.52
22. Hongqiang R, Haifei X, Liang Y, Peng L, Xiaodong L, Guanming Q (2007) Microstructure of TiB_2/B_4C composites with 1% Y_2O_3 prepared by co-precipitating and in situ synthesis. J Rare Earths 25(1):42–45. https://doi.org/10.1016/S1002-0721(07)60520-1

23. Biedunkiewicz A, Figiel P, Gabriel U, Sabara M, Lenart S (2011) Synthesis and characteristics of nanocrystalline materials in Ti, B, C and N containing system. Central Euro J Phys 9(2):417–422. https://doi.org/10.2478/s11534-010-0121-x
24. Nikzad L, Licheri R, Vaezi MR, Orrù R, Cao G (2012) Chemically and mechanically activated combustion synthesis of B_4C–TiB_2 composites. Int J Refract Metal Hard Mater 35:41–48. https://doi.org/10.1016/j.ijrmhm.2012.04.001
25. Shojaie Bahaabad M (2017) Mechanically activated combustion synthesis of B4C–TiB_2 nanocomposite powder. J Adv Mater Process 5(1):13–21
26. Azatyan TS, Mal'stev VM, Merzhanov AG, Seleznev VA (1977) Spectral-optical investigation of the mechanism of the combustion of mixtures of titanium and carbon. Combust, Explos, Shock Waves 13:156–158
27. Fan QC, Chai HF, Jin ZH (2002) Effects of particle size of reactant on characteristics of combustion synthesis of TiC-Fe cermet. J Mater Sci 37:2251–2257
28. Zhang EL, Zeng SY, Yang B, Li QC, Ma MZ (1999) A study on the kinetic process of reaction synthesis of TiC: Part I. Exp Res Theor Model, Metall Mater Trans A 30:1147–1151
29. Yang YF, Wang HY, Zhao RY, Liang YH, Zhan L, Jiang QC (2008) Effects of C particle size on the ignition and combustion characteristics of the SHS reaction in the 20 wt.% Ni–Ti–C system. J Alloy Compd 460:276–282
30. Alkan M, Sonmez MS, Derin B, Yucel O (2012) Effect of initial composition on boron carbide production by SHS process followed by acid leaching. Solid State Sci. https://doi.org/10.1016/j.solidstatesciences.2012.07.004
31. Turan A, Bugdayci M, Yucel O (2015) Self-propagating high temperature synthesis of TiB_2. High Temp Mater Proc https://doi.org/10.1515/htmp-2014-0021
32. **İpekci M, Acar S, Elmadağlı M, Hennicke J, Balcı Ö, Somer M (2017) Production of TiB_2** by SHS and HCl leaching at different temperatures: Characterization and investigation of sintering behavior by SPS. Ceram Int 43(2):2039–2045. https://doi.org/10.1016/j.ceramint.2016.10.174

Author Index

B. Li et al. (eds.), *Advances in Powder and Ceramic Materials Science 2023*, The Minerals, Metals & Materials Series, https://doi.org/10.1007/978-3-031-22622-9

Subject Index

B. Li et al. (eds.), *Advances in Powder and Ceramic Materials Science 2023*, The Minerals, Metals & Materials Series, https://doi.org/10.1007/978-3-031-22622-9